systems engineering series

series editor
A. Terry Bahill, University of Arizona

Engineering Modeling and Design

William L. Chapman
A. Terry Bahill
A. Wayne Wymore

CRC Press
Taylor & Francis Group
Boca Raton London New York

CRC Press is an imprint of the
Taylor & Francis Group, an **informa** business

CRC Press
Taylor & Francis Group
6000 Broken Sound Parkway NW, Suite 300
Boca Raton, FL 33487-2742

First issued in paperback 2019

© 1992 by Taylor & Francis Group, LLC
CRC Press is an imprint of Taylor & Francis Group, an Informa business

No claim to original U.S. Government works

ISBN-13: 978-0-8493-8011-2 (hbk)
ISBN-13: 978-0-367-40269-3 (pbk)

Library of Congress Cataloging-in-Publication Data

Chapman, William L. (William Luther)
    Engineering modeling and design / by William L. Chapman, A. Terry Bahill,
and A. Wayne Wymore.
       p. cm.
    Includes bibliographical references and index.
    ISBN 0-8493-8011-1
    1. Systems engineering. 2. Concurrent engineering. 3. Engineering models.
I. Bahill, Terry. II. Wymore, A. Wayne. III. Title.
TA168.C45 1992
620′.001′1—dc20
                                                 92-13209
                                                     CIP

Visit the CRC Press Web site at www.crcpress.com

Library of Congress Card Number 92-13209

Visit the Taylor & Francis Web site at
http://www.taylorandfrancis.com

and the CRC Press Web site at
http://www.crcpress.com

Contents

Preface

This book was written to help engineering students learn systematic principles for designing systems. We think an effective approach is with case studies; so we base the book on two, one from engineering and one from Boy Scouting.

This book was designed to be a text for an upper-division course on Concurrent Engineering or Total Quality Management, or a capstone Senior Engineering Design course. However, its basic treatment of the key issues would also make it suitable as a supplemental text for a graduate course in Systems Engineering, or as a text for a systems engineering course for industry. This book has no prerequisites, although familiarity with set theoretic notation as presented at the high school level would be helpful to the reader. The need for prerequisite material has been minimized so that this book may be used as a text for a first course in engineering fundamentals at schools with exceptional students. In the last 30 years, we have found that students who learned this material as freshmen and sophomores had an advantage over other students throughout their academic careers. This book could be used by students of all disciplines (e.g., engineering, business, sociology), except for the case study in Chapter 6, which was designed for upper-division engineering students; for this chapter, other students may need thoughtful guidance from their instructor. At innovative institutions the text could be used for a multi-disciplinary course in design. Most of the projects herein are intended for teams of students; for example, one electrical engineer, one mechanical engineer, one systems engineer, one business major, and one psychologist comprising a multi-disciplined team would be ideal.

Designing big or complex systems requires the cooperation of many people and usually many companies. In most big companies such design projects are coordinated by a Systems Engineering Department. One of the most important tasks of this department is concurrent engineering, which requires that at the very start of the project all players (e.g., customers, sales, marketing, human factors, purchasing, quality control, engineering, manufacturing, maintenance, support, repair) be involved and all facets of the system life cycle (e.g., requirements specification, testing, operation, retirement) be considered. Documenting the contributions of the concurrent engineering design team is one of the most important undertakings of systems engineering. Our two case studies are examples of such engineering documentation.

This is not a book about computer-aided design or computer-aided manufacturing (CAD/CAM). Many books and commercially available software packages already exist for computer-aided design and engineering—all specific to a particular domain; for example, there are programs for designing VLSI circuits, others for automobile suspension systems, and yet others for ten-story, steel-frame buildings. The tools presented in this book, however, are useful in all domains. The particular computer-aided tools most appropriate for the domain of interest should be used in conjunction with the general tools presented in this book.

Many engineering textbooks have chapters on probability, statistics, economics, simulation, operations research, computer tools (including database systems and hypertext), reliability, decision analysis, project management, and human factors. This book does not, because we think these topics are too important to be addressed in just one chapter. Engineering students need whole courses on each of these topics.

The first four chapters of this book present fundamental principles for designing effective systems. Chapters 5 and 6 show examples of this systems design approach. Among the many ways to use this book are the following: (1) Start at the beginning and proceed chapter by chapter, or (2) first read Documents 1 and 2 of Chapters 5 and 6, and then read the whole book sequentially. (3) Start with the first case study (Chapter 5) and bring in the theoretical material only as it is needed, or (4) present the two case studies in parallel, comparing and contrasting the analysis needed for a system composed primarily of people (a service industry) to one composed primarily of parts (a consumer product).

Chapter 7 presents other system design tools, such as the Japanese quality function deployment (QFD) (also called House of Quality and Voice of the Customer), Integrated Definition (IDEF), Design for Manufacturability, and

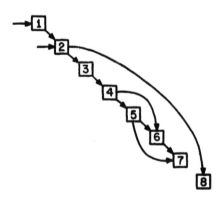

Chapter dependency chart.

Concurrent Engineering, that are often used in the system design process. We confine ourselves to tools we think will be enduring in influence or usefulness.

Because of the wide audience this book was designed to reach, a lot of the teaching comes in the projects and homework problems. Therefore, an extensive instructor's manual is available from the publishers.

We thank William J. Karnavas for his elegant sensitivity analysis of our Pinewood Derby model. This analysis helped us find and remove many design errors.

William L. Chapman
A. Terry Bahill
A. Wayne Wymore
Tucson, Arizona

Authors

William L. Chapman is a Senior Staff Engineer with Hughes Aircraft Company in Tucson, Arizona. He has worked for Hughes Aircraft since 1979 in a variety of areas, including PWB manufacturing, CAD/CAM, engineering, and electronic data systems. He is currently on staff to the development division promoting Total Quality Management (TQM) tools, such as QFD, DOE, and SPC. He received his M.S. in Systems Engineering from the University of Arizona and is currently a Ph.D. candidate in Systems and Industrial Engineering.

A. Terry Bahill has been a Professor of Systems and Industrial Engineering at the University of Arizona in Tucson since 1984. He received his Ph.D. in Electrical Engineering and Computer Science from the University of California, Berkeley, in 1975. He is a Fellow of the Institute of Electrical and Electronic Engineers (IEEE). His research interests are in the fields of modeling physiological systems, eye-hand-head coordination, validating expert systems, concurrent engineering, and systems design theory. He is the author of *Bioengineering: Biomedical, Medical, and Clinical Engineering*, Prentice-Hall, 1981; *Keep Your Eye on the Ball: The Science and Folklore of Baseball* (with Bob Watts), W.H. Freeman, 1990; *Verifying and Validating Personal Computer–Based Expert Systems*, Prentice-Hall, 1991; and *Linear Systems Theory* (with F. Szidarovszky), CRC Press, 1992.

A. Wayne Wymore is the Principal Systems Engineer of SANDS (Systems Analysis and Design Systems) and Professor Emeritus of Systems and Industrial Engineering at the University of Arizona. He received a B.S. and M.S. in mathematics from Iowa State University in 1949 and 1950, respectively, and a Ph.D. in mathematics from the University of Wisconsin, Madison, in 1955. He was the first Chairman of the Department of Systems Engineering and the first Director of the Computer Center at the University of Arizona. Dr. Wymore has consulted nationally and internationally for many government agencies and private corporations. He is the author of *Mathematical Theory of Systems Engineering: The Elements*, John Wiley & Sons, 1967; *Systems Engineering Methodology for Interdisciplinary Teams*, John Wiley & Sons, 1976; and *Model-Based Systems Engineering*, CRC Press, in press.

engineering modeling and design

chapter one

Systems engineering

The design of big or complex systems requires the cooperation of many people and usually many companies. In most big companies, such design projects are coordinated by the Systems Engineering Department. This department, which has overall responsibility for ensuring that systems designed by the company do what they were intended to do, typically includes electrical engineers, mechanical engineers, business majors, and communications specialists. The Systems Engineering Department must ensure that at the very start of the project and throughout the entire life of the system all players (e.g., customers, marketing, finance, purchasing, suppliers, engineering, manufacturing, testing, and field support) are involved and that all facets of the system life cycle (e.g., requirements specification, concept exploration, and replacement) are considered. This description is often called concurrent engineering. Systems engineering, as presented in this book, is a superset of concurrent engineering. In this chapter we discuss systems engineering, and in Chapter 7 we will examine specific attributes associated with concurrent engineering. Total quality management is an important new term in American industry. Systems engineering, as presented in this book, is a subset of total quality management. Some other components of total quality management are mentioned in Chapter 7.

Systems engineering usually is one of the first courses taken by engineering students. After learning how to design systems, they become specialists in electrical, mechanical, or biomedical engineering. The same sequence is followed in industry and senior design courses, with systems engineering first and electrical, mechanical, and other engineering disciplines covered later.

To understand what a Systems Engineering Departments does, we must first define systems engineering. This is not as easy as it sounds, because systems engineering means different things to different people. In the following paragraphs are some definitions that can be read rapidly. The differences in detail between them are not important; their similarities should be noted.

(1) Systems Engineering is concerned with the design, modeling, and analysis of technological systems that use people and machines, software and hardware, material, and energy for such purposes as communication, health care, transportation, and manufacturing. Research emphasizes tools

for modeling and analysis—especially appropriate for large, complex systems—such as concurrent engineering, system theory, decision analysis, and simulation (from a University of Arizona, Systems and Industrial Engineering brochure).

(2) Wayne Wymore, who founded the first academic department of systems engineering at the University of Arizona in 1961, says, "The principal top level function of systems engineering is to ensure that the system satisfies its requirements throughout its life cycle. Everything else follows from this function."

(3) John G. Truxal, former Dean of Engineering at Brooklyn Polytechnic Institute, says, "Systems engineering includes two parts: modeling, in which each element of the system and the criterion for measuring performance are described; and optimization, in which adjustable elements are set at values that give the best possible performance."

(4) Jaroslav Jirasec, a world-renowned Czechoslovakian systems scientist and co-founder of the International Institute for Applied Systems Analysis (IIASA), uses the Battle of Borodino from Tolstoy's *War and Peace* as his first lecture for his Systems Science class. He thinks the principles used and not used by the rival commanders Napoleon and Kutuzov and their generals define systems engineering.

(5) MIL-STD-499A defines systems engineering in the Department of Defense context. "Systems Engineering is the application of scientific and engineering efforts to (*a*) transform an operational need into a description of system performance parameters and a system configuration using an iterative process of definition, synthesis, analysis, design, test, and evaluation; (*b*) integrate related technical parameters and ensure compatibility of all physical, functional, and program interfaces in a manner that optimizes the total system definition and design; and (*c*) integrate reliability, maintainability, safety, survivability, human, and other such factors into the total engineering effort to meet cost, schedule, and technical performance objectives."

(6) A major Department of Defense aerospace contractor, Hughes Aircraft Company, starts its definition with the above three points, but adds "(*d*) verify that the hardware/software units meet their design requirements and the operational needs of the customer through the implementation of an integrated test program; and (*e*) assure the system meets its design requirements throughout the manufacture and operational life cycle of the system."

(7) A division manager at Hughes Aircraft Company defined systems engineering as performing: (*a*) requirements definition, (*b*) conceptual design, (*c*) partitioning of a system into subsystems (guidance, propulsion, etc.) for other engineering teams to create, and (*d*) system validation, i.e., ensuring the system works when the subsystems are put together to form the system. Particular attention must be paid to the interface between the subsystems.

(8) Martin Marietta, another Department of Defense and National Aeronautics and Space Administration (NASA) aerospace contractor, says that

systems engineering must ensure delivery of a system, optimized to satisfy mission requirements, that has the greatest probability of success and lowest cost. It must tie the total system together through each of the phases of the system life cycle. Thus, systems engineering can be viewed as the technical arm of program management.

(9) A draft version of MIL-STD-499B, which will replace MIL-STD-499A that was written in 1969, contains the following definition: Systems engineering is the management and technical process that controls all engineering activities throughout the life cycle in order to achieve an optimum balance of all system elements to ensure satisfaction of system requirements while providing the highest degree of mission success. It has two main activities: (a) interpreting the customer's needs and translating them into a set of requirements that can be met by individual design and speciality disciplines and (b) validating that the system satisfies the customer's needs through analysis, simulation and testing. Although originally created to help with the development of complex weapons systems, the elements of systems engineering are applicable at all design levels and to all business applications.

(10) *Systems Engineering Manual 1-1*, prepared by the U.S. Air Force Aeronautical Systems Division, defines it this way: The systems engineering process is the integrated sequence of activities and decisions that transforms a defined need into an operational, life-cycle–optimized system that achieves an integrated and optimal balance of its components. The systems engineering process produces initial, intermediate, and final products (data, equipment, trade study reports, plans, etc.) that document progress. The main tasks of systems engineering are: (a) Requirements analysis. The definition and refinement of all customer needs in terms of performance requirements and primary functions that must be performed. (b) Functional analysis. Functional analysis can be performed top down by decomposing primary functions into lower and lower subfunctions. This functional decomposition continues until each subfunction can be accomplished by a definable element. (c) Allocation. The assignment of performance requirements to the subfunctions defined by the functional analysis. (d) Synthesis. Synthesis identifies the subfunctions that must be resolved by the same element and then assigns their performance requirements to that element. (e) Trade-off studies. Trade-offs among stated customer needs shall be identified and assessed. Trade-off studies formulate and assess alternative courses of action to achieve an integrated and optimally balanced system solution.

(11) The University of Arizona catalogue says that systems engineers design and build systems to meet the needs of people. As computing speed and analytic sophistication have increased, society's needs have become more varied and complex. Graduates of the Systems Engineering program are prepared to face these needs. The goal of a systems engineer is to make the best use of resources. Stated formally, systems engineering is concerned with the processes and methodology of modeling, analyzing, and designing

technologically advanced systems that function safely, effectively, and economically. It requires appreciation and understanding of nature, machines, people, software, hardware, materials, and energy. Systems engineers work on a wide range of activities and applications, including communication systems, computer networks, manufacturing systems, robotics, health care systems, societal problems, and all phases of both industrial and military research and design. To prepare students for careers of such exceptional diversity, the Systems Engineering curriculum includes courses in computing, probability, statistics, numerical methods, operations research, concurrent engineering, knowledge-based systems, robotics, and human-system integration. This is clearly a broader and more abstract program than most traditional engineering disciplines

(12) A Case Western Reserve University brochure says, "Systems Engineering...emphasizes that problems be attacked holistically to ensure a balanced treatment of all components and their interactions....It focuses on both the theory and methodology for the analysis and design of complex technological systems and their interactions with society, economic, and environmental systems....The standard tools used in systems engineering are techniques from system modeling, optimization, decision analysis, engineering economics, control engineering, mathematics, and statistics....The systems engineering approach emphasizes critical thinking and problem solving and unifies these aspects in a mathematically rigorous framework."

(13) A U.S. Naval Academy brochure states," Systems Engineering has many definitions limited only by the number of people who attempt to define it....Systems Engineering is concerned with the design and analysis of the whole system to achieve optimum results....Since most modern complex systems are feedback in nature, the Systems Engineering program at the U.S. Naval Academy emphasizes and focuses special attention on systems engineering aspects in the light of control theory....The control engineer is placed in a unique position for solving complex problems due to his knowledge of electrical and mechanical systems and his specialized knowledge of modeling, simulation, computers, and control theory."

(14) A University of Virginia brochure states, "Systems Engineering is the intellectual, academic, and professional discipline concerned primarily with improving processes of *problem-solving* and *decision-making* throughout the life cycle of large-scale, complex man/machine/software systems in both private and public sectors."

(15) IBM says, "Systems Engineering is the iterative but controlled process in which user needs are understood and evolved—through increasingly detailed levels of requirement specification and system design—to an operational system."

(16) The following is from a syllabus for a course taught by Dr. A. Terry Bahill, University of Arizona. "When an engineering project becomes too big, too complicated, or too long-lived for one Chief Engineer to keep all the details

in his head, then responsibility for the project must be spread out amongst three or more engineers. This creates a new problem of having these engineers work independently while still being sure that their individual components will interact correctly. Ensuring that components made by different people or companies can function together is one purpose of systems engineering. A second purpose of systems engineering is to ensure that the problem is stated correctly and that tests are defined to ensure that the final system satisfies the original requirements."

We sent questionnaires to 650 alumni who received degrees in Systems Engineering at the University of Arizona over the past 30 years. We asked them what they thought systems engineers did, and this is a consensus of their responses: Systems engineers are involved in all phases of the system life cycle. They translate the customer's business needs into system requirements, evaluate alternative designs, design and evaluate prototypes, specify system testing, decompose functions into subfunctions, allocate subfunctions to physical components, analyze performance, and are involved in maintenance and operation. Systems engineers do modeling, simulation, analyses, and a lot of information gathering, writing, and planning. The following words appeared in many of their statements: multidisciplined, interdisciplinary, divergent, wide variety, interviewing the customer, communications skills, coordinate, top-down design, integrate, interface, model, trade studies, optimize, and test. Finally, several alumni expressed this sentiment: What do systems engineers do? Just about anything they want to do.

One part of this alumni questionnaire asked, "In which courses did you learn the tools, concepts, and skills that you now consider the most important?" No one course or group of courses was consistently rated the most useful. However, the courses that received the most votes were those that used computers, systems modeling courses, and project courses with written and oral reports. There was no consensus for the question, "In which courses did you learn the material that you now consider the least important?" Almost every course in and out of the department got votes. In response to the question, "What should we have taught you that we did not, or what should we have taught more of?", our alumni said that they would have liked to have had more communications skills, laboratories, computer usage, design projects, and systems modeling. The question that generated the most interesting responses was, "Who were the best teachers you had?" Almost everyone in our department got votes! In particular, two professors who normally got very low student evaluations got high marks from the alumni. This questionnaire showed that systems engineers work in diverse fields and perform a wide variety of tasks.

An important concept that can be gleaned from the definitions in this chapter is that systems engineering is responsible for the "big picture." Systems engineers must coordinate full-scale engineering, manufacturing, and component acquisition. They must design tests and evaluate proposed engineering changes during the operation phase. Finally, they are responsible for writing proposals and specifications during the design and replacement phase of the system life cycle.

Many professional groups and societies are writing standards and trying to derive a consensus on what systems engineering is and what systems engineers do. Some of these are: IEEE Systems Engineering Industry Standards Committee; IEEE Systems, Man, and Cybernetics Society; Electronic Industry Association (EIA) G-47 Committee; National Council on Systems Engineering (NCOSE); Worldwide International Systems Institutions Network (WISINET); Defense Systems Management College; U.S. Air Force Systems Command; U.S. Air Force Aeronautical Systems Division Directorate of Systems Engineering; and Department of Defense (DoD) Production and Logistics Branch.

Each of these definitions highlights different aspects of systems engineering. Taken as a whole, they might encompass systems engineering, but as the definitions themselves imply, there is no unique way to do systems engineering. In this book we present one generic approach. No company does it exactly this way, but we believe that it will be easier for the reader to understand each company's approach having first encountered this general approach.

Problems

1. **Systems engineering principles illustrated in *War and Peace*.** Jaroslav Jirasec—a world renowned Czechoslovakian systems scientist and co-founder of IIASA, a multinational systems science think tank in Austria— told us that his first lecture to his Systems Science class is based on the Battle of Borodino from Tolstoy's *War and Peace*.

For the first homework assignment, read the sections of *War and Peace* that cover August 25 and 26, 1812 (in some books they are labeled Sections 19 to 39 of Part 2 of Book III), and point out the good and bad systems engineering principles used by the rival commanders Napoleon and Kutuzov and their generals.

2. **Defining systems engineering.** Provide a consensus of the various definitions of systems engineering that were given in this chapter.

3. **The expedient engineer.** Once upon a time a mathematics student, a physics student, and an engineering student were bragging about the quality of their education. They decided to have a contest to determine who had been educated the best. The mathematics student proposed a challenge. He said, "I have been told that all odd numbers are prime. I will now use everything my professors have taught me to evaluate the truth of this hypothesis." He started thinking aloud like this: "1...3...5...7...9...11...13...15...17...19...21.... Whoops! No, the hypothesis is not true, because 21 is an odd number and it is not prime."

Next the physics student took up the gauntlet. He said, "I will now use everything my professors have taught me to evaluate the truth of the hypothesis that all odd numbers are prime." He started thinking aloud like this: "1... 3...5...7...9...11...13...15....15? Well that might just be experimental error...17 ...19...21. No, the hypothesis is not true."

Finally, the engineering student attacked the problem. He said, "I will now use everything my professors have taught me to evaluate the truth of the hypothesis that all odd numbers are prime." He thought aloud like this: "1... 3...5...7" and said, "Yep, it's true."

chapter two

The system design process

2.1 Introduction

The design process in the United States is a creative engineering activity. That is, however, a problem since the design itself is creative, but the process is not. The events that must take place to ensure a successful design are not creative because they are predictable. When the design process is changed from one design to the next, important and necessary parts of the process may be omitted. The result is a design that does not meet the customer's requirements.

Changes in the design after the system is in production indicate how poorly the design was planned and executed. Failure to consider all the requirements for producing a product, delivering it to the customer, and maintaining it will result in many changes in the design and cause a loss of confidence in the product by the customer. Figure 2.1 shows the trend for design changes over the system life cycle. The curve labeled Traditional engineering, which has higher change rates, is from the actual data of a U.S. car company. The lower rates, shown as the curve labeled Concurrent engineering, are from a Japanese automobile manufacturer that is gaining market share. A controlled design process from the outset gives the consumer a quality product and the company increased market share and profits.

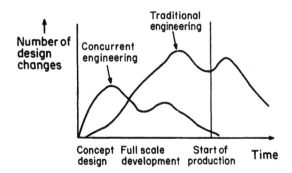

Figure 2.1 Design changes as a function of time for an American and Japanese automobile (data from the American Supplier Institute).

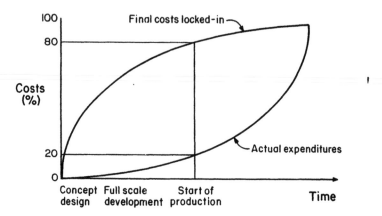

Figure 2.2 Costs as a function of time for a typical system design process.

Most costs associated with a product are determined early in its life cycle. Initial system design determines 80% of the system cost. This portion of the system can no longer be optimized with respect to cost or performance. Figure 2.2 shows that with only 20% of product development costs spent, 80% of the total product cost has already been locked in. Any future optimization of the product during its life will affect only the remaining 20% of the product's cost. By spending more time and money initially in the design cycle and ensuring that the concept selection is optimized, the company can increase the prospect of delivering a quality product to the customer.

One of the primary tasks of the systems engineer is to ensure the optimization of the design process. Systems engineering is defined as the intellectual, academic, and professional discipline principally concerned with ensuring that all requirements for a human/machine/software system are satisfied throughout the life cycle of the system. Therefore, the responsibilities of systems engineers for human/machine/software systems must also include the design and analysis of such systems. By this definition, any person responsible for documenting the requirements of a system—laying out the initial system design and ensuring that the system will do what is intended for as long as is needed—is performing a systems engineering function.

A system is broadly defined as any process or product that accepts inputs and delivers outputs; commonly encountered systems are transportation, communication, health care, food delivery, and sound (home stereo) systems. Parts of these systems are sometimes systems in themselves, such as cars, telephones, doctors' offices, restaurants, and loudspeakers. Transportation system inputs include people and goods to be transported to a destination as well as the fuel, vehicles, and other resources necessary to complete the task. The outputs are the individuals and goods delivered to a specific destination, wear on the vehicles, and pollution. The inputs for a car—a system within the

EXHIBIT 2.1

Good Systems Engineering

Seven years of systems engineering coordinating 7000 people culminated in 1969 with man's first footstep on the moon. Over the lifetime of the project six lunar modules, built at a total cost of $2 billion, landed on the moon.

The manufacturer of the lunar module, Grumman Corporation, believed that systems engineering is the design of a design—a reversal of the natural impulse to "run off and do things right away." Using NASA-supplied data on the physical properties of the moon, the design team defined system requirements, performance requirements, and test requirements; from these they developed the design specifications. They wanted to produce a system of the highest quality. Every problem that could conceivably occur anywhere in the system life cycle was considered at the beginning of the design. Later any request to deviate from a prescribed procedure had to be submitted in writing to a team of 20 project leaders, who approved it only after discussing how it might affect the system as a whole. Designing system tests was difficult. Although the lunar module was manufactured and tested on the surface of the earth, it had to be used in a vacuum with low gravity. Even with this difficulty, the system tests were specified and completed successfully. Astronauts visited Grumman often, said program manager Joseph Gavin, "to remind the Grumman crew whose neck we were protecting."

All phases of the system life cycle were considered, including retirement. After the lunar module successfully lifted the astronauts off the surface of the moon and returned them to their orbiting command module, the lunar lander was sent crashing back onto the surface of the moon as a part of a seismological experiment that made use of other measuring and recording equipment.

From Gary Stix (1988) "Moon Lander," *IEEE Spectrum*, Vol. 25, No. 11, pp. 76–82, 1988.

transportation system—include fuel, water, battery power, and the commands of the driver. Its outputs are movement, heat, and pollution.

System design often requires a knowledge of such traditional engineering subjects as physics, electrical engineering, and mechanical engineering. However, sometimes a system designer may set up a service that is a system, for

"Well, there it goes again...and we just sit here
without opposable thumbs."

Figure 2.3 System design must consider the capabilities of the user. (The Far Side cartoon
by Gary Larson is reprinted by permission of Chronicle Features, San Francisco, CA.)

which an understanding of traditional engineering is not necessary; for
example, setting up a fast food service in a restaurant requires only an
understanding of the design process and the food industry. Thus, a systems
engineer may design a system and yet have no training as a traditional
engineer.

In this chapter we look at the design process in general; the specifics of
system modeling are discussed in the ensuing chapters.

2.2   The design process

The design of any product or service must begin with a complete understanding of the customer's needs. It does no good to create a product that
appeals to engineers, but that no one else can use (see Figure 2.3). Customer
requirements must be considered by all involved in the project. The designers
need this information to design the optimal product or service.

The information necessary to begin a design will usually come from
marketing studies or specific requests from a customer. Most consumer
products begin with a detailed marketing study that determines a need for a
new product or service. If the product is large, other companies or suppliers
may be used to make parts of the total system. In this case, the requirements
for those portions of the system are usually delivered by the main company

EXHIBIT 2.2

Excellent Cooperation between Engineering and Manufacturing

In 1983 IBM released the PCjr™. It was an excellent example of cooperation between Engineering and Manufacturing. These departments worked together to produce a computer system with half the parts of its predecessor, the IBM PC. Unfortunately, Sales and Marketing were not involved in the design effort and they did not find out what the customer wanted. After the PCjr was released the customers thought it was too small—the keyboard was too little, memory was insufficient, and disk space was inadequate—and it was not expandable. In addition, a lot of the PC's software would not run on the PCjr. So, in spite of the expensive and clever advertising campaign featuring cartoons of Charlie Chaplin's character, The Tramp, the PCjr was a sales flop. The designers failed because they did not consider the wants of the customer during system design. The "Voice of the Engineer" won out over the "Voice of the Customer."

For further reading, see John Voelcker, Paul Wallich, and Glenn Zorpette (1986) "Personal Computers: Lessons Learned," *IEEE Spectrum*, Vol. 23, No. 5, pp. 44–75.

as a combination of specifications, drawings, and verbal requests. This gives the supplier a good understanding of the main company's needs; so the resultant risk should be less. It is the responsibility of the systems engineer to ensure that all information on the customer's requirements is collected and made available for the purpose of system design.

After the decision to design a system has been made, a design manager is appointed to oversee the project. The design manager assembles a team of specialists to create the system design. He first selects one or more systems engineers to collect and document all the system requirements. Direct interviews with the customer are necessary to ensure that the words used have the same meaning for everyone involved. For example, the customer may say a liquid should taste hot, and this might be translated into a requirement for a temperature from 110 to 115 °F. It is important to keep accurate minutes of all meetings with the customer so as not to overlook any details and so that all requirements are traceable to the customer.

Any "stakeholder" of the system is a customer of systems engineering. In the design of a horse racing system, the customers include the track owner, the bettor, the jockey, and the horse! All of these have an interest in the system,

EXHIBIT 2.3

Human System Integration

Humans operate most manufactured systems, and if their tasks are considered at an early stage, productivity can be greatly enhanced.

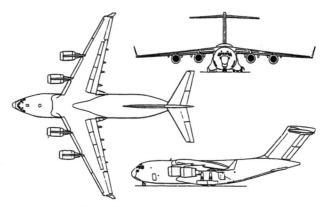

Figure 2.4   C-17A long-range heavy lift cargo transport.

The Air Force C-17-A long-range, heavy-lift, cargo-transport airplane originally needed three ground workers to refuel it: two people pumped the fuel and one sat in the cockpit to operate the fuel boost pump switches. In an emergency the ground crew depended on the person in the cockpit to turn off the switches quickly.

After studying the operation of these three-person crews, the designers moved the switches to the airplane's wheelwell. Two people could now refuel the aircraft, and in case of an emergency, they could turn off the switches a lot faster than could the person in the cockpit. The redesign increased safety and decreased human resource requirements. If the actions of the refueling team had been considered early in the design process, this change would not have been necessary.

For further information contact IMPACTS Office, HQ USAF/PRME, Pentagon, Washington, D.C. 20332.

and all can influence the performance of the system. If the horses are not fed they will not perform well; likewise, if the bettors are not comfortable they may go home early.

Members of an engineering design team may include mechanical engineers, electrical engineers, manufacturing engineers, and logistics, finance, marketing, and procurement personnel. Including all players early in the design process is the basis of concurrent engineering. For example, the team that designs a health care service may consist of a doctor, nurse, health care administrator, billing clerk, and suppliers. Initially, their job is to provide the systems engineer with advice on the correct way to express the requirements for the design. The customer of the health care system is the patient who does not feel well and wants medical attention. The requirement of this customer seems fairly simple: to feel better after leaving the doctor's office. This may be the major concern of the patient, but some other concerns are the cost of the service, the length of wait for service, the cleanliness of the waiting room, and the attitudes of the workers. These concerns must somehow be considered and quantified so that the customer can be satisfied. The doctors, nurses, and other design team members know their individual specialties and usually have a good idea of what information is needed. How long the patient is willing to wait before being seen may depend on how bad the patient feels. The patient may need to be categorized in terms of how urgently care is needed. A nurse can judge whether a patient should first fill out the paper work or be seen by the doctor. The acceptable waiting time could be based on an urgent care index provided by the nurse.

Every requirement defined by the customer or the specialist must be testable. This is an axiom of systems engineering: "It isn't a requirement if it isn't testable." An agreed upon way of measuring the performance of the system is necessary to ensure that the system does what it is supposed to do. If the customer wants the liquid hot but its temperature is never tested, then at some time in the future the customer will receive a product that does not meet the requirements. Sometimes customers may not want to test the product or the service to the extent recommended by the systems engineer. Still, the systems engineer must specify how the system ought to be tested so that the test requirement is complete and no requirement is overlooked. Customers will often change their minds and then want the test ability.

An example of a failure in system testing was the Hubble space telescope (see Exhibit 2.4). The design and construction of this $1.5 billion device spanned a decade, yet a final system test was never done to verify that it worked. After the telescope was deployed in space, it was discovered that the optical mirrors had been ground incorrectly. The problem was traced back to a measuring device used in polishing the mirrors. The requirement for superior accuracy was in the specifications, but the alignment device was never properly tested or controlled. The engineers just assumed it would work, with disastrous results.

The next step in the design process is to make certain that the requirements of the customer can be met. This is called requirements validation and is an important step. If no system can be built to satisfy all the customer demands, there is no point in proceeding. For example, if a customer requests a 2000 pound automobile that gets 1000 miles per gallon of gasoline without any other energy source, either the marketing department or the company

EXHIBIT 2.4

Lack of System Test

The Hubble Space Telescope was launched April 24, 1990, after over a decade of designing, manufacturing, and testing that cost $1.5 billion. Freedom from atmospheric disturbances and high precision manufacturing was supposed to give it ten times the resolution of ground-based telescopes. To achieve this resolution, the primary mirror was ground with a precision to better than 150 Å. But it was ground in the wrong shape! This produced a telescope with one-third the specified resolution that was incapable of resolving faint objects, such as planets of nearby stars.

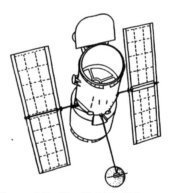

Figure 2.5 The Hubble Telescope.

During the telescope's manufacture there were many clues that something was wrong. In fact, a photograph taken in 1981 shows the flaw, but test procedures directed the tester's attention to a different part of the photograph. Each component worked very well by itself; the failure occurred when the components were put together to form a telescope. The failure would have been detected if a full system test was performed; but alas, such a test was not required. Tests were specified to detect small errors but not blunders of this magnitude. A complete system test was thought unnecessary and too costly; however, the cost of a complete system test would have been much less than the $50 million estimated cost of fixing the telescope in orbit.

For further reading, see George Field and Donald Goldsmith (1989) *The Space Telescope*, Contemporary Books, Chicago, and Richard T. Fienberg (1990) "Space Telescope; Picking Up the Pieces," *Sky & Telescope*, Vol. 80, No. 4, pp. 352–358.

requesting the project must be notified that a system cannot be built to satisfy the requirements. Further effort would be a waste of time and money. Sometimes the validation can be very difficult. Meeting the cost requirement and obtaining adequate functionality may seem possible with a new technology, but there is a risk involved in validating the requirements, which should then be made obvious in the documentation. On the other hand, sometimes validation is easy, it being sufficient to merely show that an existing system meets the customer's requirements. For example, if a company decides to enter the television market and a competitor has a product on the market that fully satisfies the consumer, then the validation of the customer's requirements is obvious. The company's requirements for profit or market share must be validated on their own merits.

Once the design requirements have been detailed and validated, the design manager will select team members to begin concept exploration. This team should include the original team members, but this is not always necessary or possible. A brainstorming session is an opportunity for the creative thought process to take place, where all team members participate and discuss ideas. No ideas should be excluded, since some of the most outrageous ideas can become the technical breakthroughs of the future. The systems engineer is responsible for documenting the session and ensuring that the discussion addresses the customer's requirements. A systems engineer with a broad background can act as referee for the many ideas that will be presented.

After a set amount of time, the design manager should select the most appealing choices—as many as can be handled with the time and budget allotted—and have the team produce trade studies. Trade studies are research documents that compare the performance and costs of alternative designs. Literature searches, past experience, and analyses of models can be used to produce the trade study. Most designs require a model upon which analysis can be focused. The analysis should include a measure of all the characteristics the customer wants in the finished product. The concept selection will be based on the measurements from the models.

The models are created by first partitioning each conceptual design into functions. This decomposition often occurs at the same time major physical components are selected. For example, the design of a new car requires either mechanical or electronic ignition systems. These are two separate concepts. The top level function, firing the spark plugs, is the same, but when the physical components are considered, the functions break out differently. The firing of the spark plugs is directed by a microprocessor in one design and a camshaft in the other. Both perform the same function, but with different devices. Which is superior will be determined by customer requirements such as cost, performance, and reliability. These characteristics are measured based on the test criteria. System analysis and trade-off studies are done by the systems engineer. Other members of the team will help build the models depending on the technology being used.

Building the models is of primary importance. Designing all the customer's desires into the product is critical to satisfying the business goals of the company. For this reason, a large portion of this book is devoted to a system modeling language. The mathematical and technical skills of the systems engineer are critical during this phase. Accurate simulations and optimizations of a system's design and performance will result in the selection of the best concept.

This may best be illustrated by an example. The space shuttle is an expensive, highly complex system. Before the present configuration was built, all the elements of its design were modeled in detail. In this manner, the size of the engines, cargo bay, and launch pad were all optimized. The size of the cargo bay, in turn, had a large impact on what the shuttle carries. The Hubble space telescope was designed to fit into the cargo bay. If the bay had been larger the telescope would also have been larger. The trade-offs in building the shuttle were carefully modeled before any component was created to try to predict the impact it would have on other systems such as the telescope.

Once the concept selection has been made, the entire system functional analysis and a physical synthesis of the concept must be done. Careful attention to the satisfaction of all system requirements must be given. The system test requirement is used throughout this stage to ensure a valid system design. The main purpose of the physical synthesis is to prove that the system can be built, not to optimize every element. That process is reserved for the full-scale engineering of the product and processes.

Detailed design begins once the design of the system is complete. The members of the team will do the separate design jobs. The systems engineer is responsible for seeing that the requirements are still met during full-scale engineering. Design reviews are held throughout this phase to verify that all requirements are met and that the interfaces between the different components of the system are correct. The design should not proceed unless a consensus on the design has been reached during design reviews.

This applies not only to engineering systems, but also to services. For example, in a fast food restaurant, the person who takes the order must have supplies nearby to deliver the food. If the food delivery component of the system was designed to deliver large bottles of ketchup and the order counter was designed to dispense small packets, there would clearly be a problem. The interface between the server and supplier components should have been better defined so that the supplier knew in advance the requirements of the server. The requirements for interfacing the supplier and server components should have been checked (or validated) through a design review.

The design review is a formal meeting that brings experts from various disciplines together to review the design package. Many reviews are held throughout the system life cycle. In the review, the designers present their work with any analysis necessary to show why they selected the approach, components, and suppliers they are using and whether the system meets all the system requirements, which include function, cost, and schedule. Most errors are caught before the design review, but many important decisions are changed and mistakes prevented through the formal review process.

The system enters production after the design stage. The transition to production is error prone. Regardless of the quality of the drawings, mistakes in interpretation and intent will be made. Concurrent engineering requires that the processes are designed at the same time as the product. This will limit the number of mistakes made during the transition to production. For example, a designer may use a new plastic for a console, but manufacturing has found this material hard to work with. Concurrent engineering would catch

EXHIBIT 2.5

Catastrophic Design Errors

In 1986, the General Electric Company (GE) marketed a new refrigerator. Their engineers were confident that their new compressor would help them leapfrog the Japanese; yet, in 1988 they declared a loss of $450 million. What went wrong?

The story began in 1981, when the market share and profits of the GE Appliance Division were falling. GE was using 1950s technology to make compressors. Each compressor took 65 minutes to make, whereas Italian and Japanese companies made theirs in 25 minutes, with lower labor rates.

The GE engineers said they could reduce the part count by one-third by replacing the reciprocating compressor with a rotary compressor, like the one used in their air conditioners since 1957. Furthermore, they said they could make it easier to machine by using powdered-metal instead of steel and cast iron for two parts, thereby cutting manufacturing costs. Although powdered-metal had failed in GE air conditioners a decade earlier, no one on the new design team had experience with this previous failure, and evidently, they felt no need for advice from people involved in a failed project. They also turned down advice from Japanese and American consultants with experience in designing rotary compressors.

Six hundred compressors were "life tested" by running them continuously for two months under temperatures and pressures that were supposed to simulate five years of actual use. Not a single compressor failed, and the good news was passed up the management ladder. (Doesn't this sound too good to be true?) During testing, technicians noticed that many of the motor windings were discolored from heat, bearing surfaces appeared worn, and the sealed lubricating oil seemed to be breaking down. This bad news was not passed up the management ladder!

GE offered a five year warranty on the refrigerators, but they could not wait five years before beginning full-scale manufacturing. Evaluating a five-year life span based on two months of testing is tricky, so the original test plan was to field-test some refrigerators for two years before full-scale manufacturing began. Pressure to stay on schedule reduced this test time to nine months.

By the end of 1986, GE had produced over one million new compressors. Everyone was ecstatic over the new refrigerators; however, in July of 1987 the first refrigerator failed. Quickly thereafter came an avalanche of failures, and the engineers could not fix the problems. In December of 1987, GE started buying foreign compressors for the refrigerators. Finally, in the summer of 1988 the engineers made their report. The two powdered-metal parts were wearing excessively, increasing friction, burning up the oil, and causing the compressors to fail. GE management decided to redesign the compressor without the powdered-metal parts, and in 1989 they voluntarily replaced over one million defective compressors.

The designers who specified powdered-metal made a mistake, but everyone makes mistakes. Systems engineering is supposed to reveal such problems early in the design cycle or at least in the testing phase. This example was a failure of systems engineering.

For further information, see Thomas F. O'Boyle (1990) "GE Refrigerator Woes Illustrate the Hazards in Changing a Product," *Wall Street Journal*, 7 May, Section A.

this problem during the design phase. The traditional approach of implementing the manufacturing processes after the design is complete would require a design change and a delay in implementation. The system requirements must be strictly followed during production. If the original requirements fully allowed for manufacturing difficulties, there should be no need for design changes.

The systems engineer is responsible for the system test. This test verifies that all components work when put together on the factory floor, in the field, or at a construction site.

The systems engineer's responsibility does not end when the product or process is in service. The operational support, maintenance, and retirement of a product must have been planned as original requirements; however, unforeseen events occur and enhancements are common, so it is necessary for the systems engineer to be available to support systems in operation. If the project has been properly documented throughout its design, engineering support should be a direct and inexpensive activity or task.

2.3 System design

In this section we formally describe the elements of a system design.

2.3.1 System requirements

There are six categories of system requirements that the systems engineer must specify:

1. Input/Output and Functional Requirement
2. Technology Requirement
3. Input/Output Performance Requirement
4. Utilization of Resources Requirement
5. Trade-Off Requirement
6. System Test Requirement

The Input/Output and Functional Requirement consists of definitions of the time scale, the set of all admissible inputs over time, the set of all eligible outputs over time, and the required functional relationship between the inputs and the outputs. It represents what the system must do independent of the technology.

The Technology Requirement consists principally of limitations, as specified by the customer, on the technologies available to build the system. It can list certain components or processes that shall be used or not used to solve the problem. It may also include budgets and schedule constraints for the design.

The Input/Output Performance Requirement specifies how well the Input/Output and Functional Requirement is to be met. This may be expressed in measurable terms called figures of merit, and for any design under consideration it is necessary to be able to estimate or measure the values of these figures of merit.

The Utilization of Resources Requirement specifies how well the Technology Requirement is to be met. It is expressed in terms of figures of merit also. An example of a typical measure is operating cost per year.

The Trade-Off Requirement specifies the nature of the trade-offs between the Input/Output Performance Requirement and the Utilization of Resources Requirement. The Trade-Off Requirement will make the actual system selection based on the priorities of the customer.

The System Test Requirement specifies the methods for observing and testing the final system that is built. The System Test Requirement includes specifications for estimating or measuring values for all the figures of merit defined as part of the Input/Output Performance, Utilization of Resources, and Trade-Off Requirements. These estimates are made on the basis of "blue-sky" approximations, analyses of models, or data collected from testing the prototypes, the system models, or the final system.

2.3.2   System life cycle

The life cycle of a human/machine/software system is defined in seven phases. Systems engineering must ensure that system requirements are satisfied throughout all seven phases.

1. Requirements Development Phase
2. Concept Development Phase
3. Full Scale Engineering Development Phase
4. System Development Phase
5. Test and Integration Phase
6. Operations, Support, and Modification Phase
7. Retirement and Replacement Phase

Systems engineering reports are generated during these phases. During the first two phases, systems engineering creates the following seven systems engineering documents:

Document 1: Problem Situation
Document 2: Operational Need
Document 3: System Requirements
Document 4: System Requirements Validation
Document 5: Concept Exploration
Document 6: System Functional Analysis and Decompostion
Document 7: System Physical Synthesis

Each of the phases and documents is described in the ensuing sections of this chapter. The table of contents of the Pinewood and SIERRA case studies in Chapters 5 and 6, respectively, show the information in each of the seven documents. A simple example of this system design documentation is presented in Exhibit 2.6. A comparison of the text below with the appropriate sections of Exhibit 2.6 should prove helpful in understanding the intent of each document.

2.3.2.1   Requirements development phase

In the Requirements Development Phase, systems engineering must collect information from the customer to produce a set of requirements. These requirements are then validated to ensure their consistency. No solution is considered or addressed in this phase unless it is needed to clarify requirements. The goal of this phase is a clear and precise yet comprehensive statement of the requirements. During this phase, four of the seven engineering documents are created that describe the requirements and their validation. (Refer to

Exhibit 2.6 and the Pinewood and SIERRA case studies for examples of these documents.)

The Problem Situation Document is the executive summary. It explains the overall problem, the history of the project, who the customer and designers are, and the plan that will be used to control the design process. This document is written in plain language and is intended for management. It should be continually updated.

The Operational Need Document states in plain language what the customer expects from the new system. It is described in terms of the six categories of requirements mentioned in Section 2.3.1. The human problem, not the technical problem, that needs a solution is described in this document. It is intended for management, the customer, and systems engineers. Exhibit 2.6 has a simplistic example of an Operational Need Document.

The System Requirements Document describes mathematically, or in complete textual detail, each of the requirements addressed in the Operational Need Document. Figures of merit for each requirement are stated, along with detailed tests to measure them. The audience is systems engineers. This document should be as long as is necessary to describe the requirements completely. In Exhibit 2.6, Document 3 describes the technical requirements for making a loud sound.

The System Requirements Validation Document contains an examination of the mathematical description of the requirements presented in the System Requirements Document to check for consistency and a demonstration that a real world solution can be built and adequately tested to prove that it satisfies the requirements. Identifying an existing system that satisfies all the requirements is usually enough to complete the system validation. Risks associated with meeting the requirements should be discussed when applicable.

2.3.2.2 Concept development phase

After validating the system requirements either by finding at least one system concept that solves the problem or by system analysis, the next responsibility of systems engineering is to select the best system design concept to pursue for the design of the final system. A system design concept, in this context, is a class of solutions that all have the same general form. Systems engineering selects the best system design concept from all the perceived alternatives by comparing all the alternative system design concepts with regard to the satisfaction of system requirements. The system design concepts will guide all future system functional analysis and physical synthesis activities. The best system design concept is selected on the basis of the approximation, simulation, or measurement of the figures of merit and on a trade-off analysis.

The Concept Exploration Document identifies the best of the alternative design concepts through the use of trade studies and modeling. For each concept, a value for each figure of merit is approximated, simulated, or derived from a prototype. A trade-off and sensitivity analysis is done based

EXHIBIT 2.6

A Trivial Example of the Seven Systems Engineering Documents

Document 1: Problem Situation
> Top Level System Function: Make a loud noise.
> Customer: A teacher who wants to get attention.
> Environment: Noise to be in a normal classroom environment.

Document 2: Operational Need
> Deficiency: A teacher needs to get the attention of the students.
> Input/Output: Input is energy from the teacher; output is sound.
> Technology: The teacher is available as a human resource; a wrist-watch accurate to 1 second; and an assistant for testing.
> I/O Performance Requirement: The sound must be heard at the back of the classroom.
> Utilization of Resource Requirement: Sound must be made by the teacher soon after the decision to make the sound. The teacher prefers to put in minimal energy.
> Trade-Off: The I/O Performance Requirement will be given 75% of the points and the Utilization of Resources 25% of the points.
> System Test: The teacher will make the sound. The assistant will stand at the back of the room and record if it is heard.

Document 3: System Requirements
> System Requirement: Sound is loud enough to be heard at 30 feet.
> Input/Output
> > Input 1: Physical energy from the teacher.
> > Input 2: Time of day.
> > Output 1: Sound at 30 feet.
> Technology: The teacher is the only physical resource to build the system. A wristwatch, accurate to within 1 second and an assistant are available for testing only.
> I/O Performance Requirement: The total I/O Performance (IP) is computed using 2 points for an easily heard sound, 1 point for a barely audible sound, and 0 points for no sound.
> Utilization of Resources: The total Utilization Performance (UP) is computed using 2 points for a sound within 2 seconds and 0 points for a sound after 2 seconds, plus 3 points if the teacher has expended minimum energy, 2 points for moderate energy, and 0 points for a large amount of energy.
> Trade-Off: System score = $(0.75 \times IP) + (0.25 \times UP)$. Highest score wins.
> System Test: The assistant will stand 30 feet from the teacher, say "Go," and then time how long the teacher takes to make the sound and judge how loud the sound was. The teacher will judge how much energy was put into the system.

Document 4: System Requirements Validation
> These requirements can be satisfied by a teacher clapping within 2 seconds of making the decision to do so.

Document 5: Concept Exploration
The following data came from experiments with prototypes:

Concept 1: Teacher claps hard once. Using the test requirement the scores are:

IP score:	2 points	$\rightarrow$	Easily heard
UP score:	2 points	$\rightarrow$	Within 2 seconds
	+2 points	$\rightarrow$	Moderate energy

System score $= (0.75 \times 2) + (0.25 \times 4) = 2.5$

Concept 2: Teacher shouts "Hey!". Using the test requirement the the scores are:

IP score:	2 points	$\rightarrow$	Easily heard
UP score:	2 points	$\rightarrow$	Within 2 seconds
	+3 points	$\rightarrow$	Minimum energy

System score $= (0.75 \times 2) + (0.25 \times 5) = 2.75$

Concept 3: Teacher stomps foot hard one time. Using the test requirement the scores are:

IP score:	1 point	$\rightarrow$	Barely audible
UP score:	2 points	$\rightarrow$	Within 2 seconds
	+0 points	$\rightarrow$	Lots of energy

System score $= (0.75 \times 1) + (0.25 \times 2) = 1.25$

Recommendation: The best was Concept 2, scoring 2.75.

Rationale for Concepts: Other concepts—popping a balloon, banging on a drum, and lighting a firecracker—were all discarded early in the process because they required additional resources.

Document 6: System Functional Analysis and Decomposition
The top level system function is to make a sound loud enough to be heard at 30 feet within 2 seconds.

Concept 1: Teacher claps hard once. The functions are:
(1) Make decision, (2) Raise hands, (3) Clap hands.

Concept 2: Teacher shouts, "Hey!" The functions are:
(1) Make decision, (2) Open mouth, (3) Shout "Hey!"

Concept 3: Teacher stomps foot hard one time. Functions are:
(1) Make decision, (2) Raise foot, (3) Slam foot to the floor.

Document 7: System Physical Synthesis.
Concept 1: Teacher claps hard once. Physical unit is the teacher. Components are the teacher's brain, arms, and hands. The brain decides, the arms rise up, and the hands produce the sound when put together.

Concept 2: Teacher shouts, "Hey!" Physical unit is the teacher. Components are the teacher's brain, mouth, and voice box. The brain decides, the mouth opens, and the voice box produces the sound.

Concept 3: Teacher stomps foot hard one time. Physical unit is the teacher. Components are the teacher's brain, leg, and foot. The brain decides, the leg rises up, and the foot produces the sound when it hits the floor.

on these numbers, producing a recommendation for the best concept. This document will be rewritten many times as more information becomes available. It is intended primarily for systems engineering personnel. Refer to Exhibit 2.6 for a trivial example of concept exploration.

The process of functional analysis and physical synthesis is begun when a concept is considered worthwhile. System Functional Analysis begins with the top level system function, which is the Input/Output and Functional requirement as described in the System Requirements Document. The first step in system functional analysis is to decompose the top level system function into a related set of system functions, each defined in the same format as, and consistent with, the top level system function. This is known as top down design. The decomposition must be done with the aim of resolving the functional requirements into portions small enough to be addressed. The entire set of smaller system functions will satisfy all the functional and test requirements. The objective of this process is to define simple tasks, or system functions, that can be performed by people, machines, and software and that, when properly organized, will achieve the performance of the top level system function. In Document 6 of Exhibit 2.6 we decompose the top level function (make a loud sound) into the functions of (1) make decision, (2) raise hands, and (3) clap hands together. The combination of the three small functions forms the top function. Notice that we have selected our technology and then defined our functions with this technology in mind.

For each system function identified at a given stage, systems engineering defines a complete set of system requirements consistent with previously defined system requirements. This is called requirements decomposition and allocation. The System Functional Analysis Document is used to track the decomposition of the chosen concept into successively smaller system functions that will eventually be simple enough to implement physically. Each function is described in three ways: in plain language, graphically, and in mathematical terms. Interrelations between the smaller functions are carefully described to ensure that the overall system function is satisfied. The intended audience of the document is systems engineering personnel.

Physical Synthesis is concurrent with system functional analysis. The decomposition of the top level system function is based on guidance from physical synthesis. At each stage of the decomposition, systems engineering identifies, in increasing detail, the people, machines, and software and their interrelationships necessary to perform each identified system function. This process, physical synthesis, is distinct from system functional analysis. The first step in physical synthesis is to demonstrate the validity of each system function by identifying a system design concept for its implementation. Next, physical elements are assigned in increasing detail to each system function consistent with the design concept stated in the Concept Exploration Document. Finally, it must be demonstrated that all system functions will be performed, all system requirements will be satisfied, and all components of the physical design are justified by the system requirements. For example, in

Document 7 of Exhibit 2.6 the physical elements are described for the functions of Document 6. It is possible to have different physical elements perform the same function: a student could make the decision for the teacher.

To end the Concept Development Phase, systems engineering writes a detailed set of specifications for each one of these physical elements in such a way that hardware, software, and human factors engineers can develop a detailed design in the Full-Scale Engineering Development Phase. The System Physical Synthesis Document will describe the decomposition of the system design concept into physical units and will assign the system functions to a system physical element for implementation. Each system physical element is described in three ways: in plain language, graphically, and in exact mathematical detail. Interrelations among the system physical elements and interfaces between inputs and outputs must be documented. The intended audience of this document is systems engineering personnel.

2.3.2.3 Full scale engineering development phase
In the Full Scale Engineering Development Phase, hardware, software, and human factors engineers—coordinated by systems engineering—produce detailed specifications, drawings, and programs necessary to produce and manufacture hardware and software elements and to train operators and users. These specifications are based on requirements generated by systems engineering for each physical element, as recorded in the System Functional Analysis Document and the System Physical Synthesis Document. Design reviews are held throughout this phase to coordinate all development work.

2.3.2.4 System development phase
In Phase 4 of the System Life Cycle, System Development, systems engineering helps coordinate the efforts of industrial and manufacturing engineers who have the primary responsibility for the physical creation of the system. The basis of this effort is the system design and the detailed designs produced during the Full Scale Engineering Development phase. The systems engineer and the development engineers must assist in implementing the design in production. If concurrent engineering has been used, the production processes are already in place.

2.3.2.5 Test and integration phase
After the physical elements of the system have been produced, System Life Cycle Phase 5, Test and Integration, begins. Systems engineering oversees the integration of the physical elements into a system in which the physical components work together to satisfy the customer's operational need. Systems engineering is also responsible for testing the system and documenting the test results. The tests performed in this phase were designed and documented during Phase 1 of the System Life Cycle. It is important that manufacturing not perform the system test because of a conflict of interest; systems engineering is the customer's representative in the engineering process.

2.3.2.6   Operation, support, and modification phase

During the Operation, Support, and Modification Phase, systems engineering is responsible for periodic reevaluation of the system and for controlling changes in the system. The original system requirements must have allowed for logistic support of the system. Modifications to the design necessitate a review of the entire set of systems engineering documents. A large modification will require the entire design process to be redone. Integrated Logistics Support (ILS) fully supports a system in operation. This includes the maintenance, training, and service needed to satisfy the customers.

2.3.2.7   Retirement and replacement phase

In System Life Cycle Phase 7, Retirement and Replacement, the acquisition cycle begins again. Systems engineering is responsible for recommending the retirement of the system at the appropriate time (for a creative system retirement see Exhibit 2.1) and for writing the proposal for its replacement. During this phase, everyone is appreciative of the careful and precise documentation done during Phases 1 and 2. The rationale for the existing system design can be easily discerned from the requirements development and concept selection documentation, pointing the way for a new system design. For example, if the design of a car was carefully documented during the early phases of its life cycle, the redesign of this car would only require an update of the existing system requirements and trade studies. The time to complete the design effort would be greatly reduced. The best Japanese automobile manufacturers bring a new product to market every four years, whereas traditional U.S. companies take eight years. This is largely due to the Japanese following a good system design process.

2.4   The order in which the documents are written

> *"Begin at the beginning,"* the King said very gravely, *"and go on 'til you come to the end; then stop."* (From Lewis Carroll, *Alice's Adventures in Wonderland.*)

That may be how an event is described, but it is not how one writes system design documents, which are constantly changing and being updated. New data is added and the old removed as the project progresses. Initially, the Problem Situation and Operational Need Documents are created. A rough draft of the System Requirements Document is next created and then Concept Exploration begins, the concepts being necessary to appropriately define system requirements in detail. For example, if a trip to Mars is a manned flight, the human factors involved must be considered before a System Requirements Validation can be done. The iterative approach is to rough out some concepts, then improve the System Requirements Document and the System Requirements Validation Document. Once enough detail has been put into these documents, a more complete set of concepts can be derived. This does not imply that the requirements are written with a solution in mind, but rather, that the requirements cannot be fully detailed until a set of concepts exists. After a set of concepts is selected, the System Functional Analysis and Decomposition and System Physical Synthesis Documents can be started. Often, system requirements will be modified as system functions are better described. No systems engineer thinks of everything at the start, and bad engineering is the result of an inflexible system design. Data from Documents 6 and 7 are fed back into Document 5 until a firm decision on the best concept is made; then all documents are updated and completed.

2.5 *Summary*

The process of system design is important, requiring an orderly procedure. There is no magic for doing it right the first time; careful consideration of the customer's needs throughout the life cycle of the product will guarantee a satisfied customer and a successful design.

Problems

1. **Defining the customer.** Assume that you have just been hired as a teaching assistant for a college course in Numerical Methods. Before you start lecturing, you decide to design a system for teaching and administering this course. The set of *all* people who have the right or responsibility to impose requirements on a system is the customer. Describe your customer by providing information for each of the items in the portion of the Systems Engineering Document 1 shown below. You are limited to one page.

Owners
Bill Payers, the Client
Users
Operators
Beneficiaries
Victims
Technical Representatives
Social Impact
Economic Impact
Environmental Impact

2. **Requirements specification.** Suppose your boss wants you to order a digital clock from a company in Japan. You do not have their catalogue, so you must describe what you want so as to obtain their most acceptable item. Write the requirements that you intend to send them. Make sure you get it right, because it takes a long time to receive things from Japan and it is too expensive to clarify things on the telephone (your boss will be mad if you

waste money). You should take no more than two pages in specifying the following requirements:

Input/Output
Technology
Performance
Utilization of Resources
System Test

3. **System life cycle–1.** Assume you were hired on June 1 to teach a two-semester course on systems engineering that would run from September to May. Assign the seven phases of the system life cycle to the twelve months.

4. **System life cycle–2.** Like other terms used in this book, there is no unique definition of the system life cycle. Ask a dozen companies and you will get a dozen definitions. The definition will even differ for a product design, such as a bus, versus a project design, such as a municipal bus system. The following two life cycles come from Kerzner, 1989.

Life cycle 1
Research and development
Market introduction
Growth
Maturity
Deterioration
Death

Life cycle 2
Conceptual Phase
Definition Phase
Production Phase
Operational Phase
Divestment Phase

Which of these is for a product and which is for a project?

5. **System life cycle–3.** The definition of system life cycle even varies from industry to industry. These four life cycles come from Kerzner, 1989:

Life cycle 1
Startup
Definition
Main analysis
Termination

Life cycle 2
Formation
Buildup
Production

Phase-out
Final audit

Life cycle 3
Conceptual
Planning
Definition and design
Implementation
Conversion

Life cycle 4
Planning and data gathering
Studies and basic engineering
Major review
Detailed engineering
Construction
Testing and commissioning

Which of these life cycles is from Construction, Manufacturing, Computer Programming, and Engineering?

6. **Concept exploration.** Assume you were hired on June 1 to teach a two-semester course in systems engineering that would run from September to May. What concepts would you explore in the Concept Exploration Document that you would write in the Concept Development Phase of the system life cycle?

7. **System test.** At the behest of the federal government, many states, counties, and municipalities have mandated that gasoline contain ether or alcohol to reduce carbon monoxide emissions. If you are in one of these localities, find out what the performance requirements are (i.e., what is supposed to be improved and by how much), and find out what system tests were designed to verify air quality improvement. Be aware that these "oxy-fuels" also increase formaldehyde emissions, and aldehydes are injurious to the human respiratory system.

8. **Functional decomposition.** Assume that you will be teaching a course in systems engineering this coming semester. Do a functional decomposition of your job. Consider administrative as well as academic matters.

9. **Retirement and replacement.** What should be done with the Hubble Space Telescope when it reaches the end of its four year life cycle?

10. **Extinguishing a candle.** Write the seven systems engineering documents for a system to extinguish a candle. You are limited to two pages.

11. **Moving a sheet of paper.** Write the seven systems engineering documents for a system to move a sheet of paper (8.5×11 inches or 21×29.7 centimeters) across the room (10 feet or 3 meters). You are limited to two pages.

12. **Design philosophies.** We have been emphasizing that systems must be designed; nothing in the development should be haphazard. We have advocated the philosophy "ready, aim, fire"; however, many successful entrepreneurs seem to use the philosophy "ready, fire, aim," and companies with many MBAs on staff use the philosophy "ready, aim, aim, aim." When would *you* use each of these philosophies?

chapter three

A tool for modeling systems

3.1 What constitutes a system?

A system is any process that converts inputs to outputs. A system creates outputs based on inputs, over which it has no direct control, and the system's present state. A system commonly encountered is the traffic light controller at a street intersection. It accepts inputs, such as pedestrians pushing the walk button or cars driving over sensors, and creates outputs that are the colored lights in each direction.

3.1.1 Definition of state

Defining the state of a system is one of the most important, and often most difficult, tasks in system design. The state of the system is the most concise description of its past history. The current system state and a sequence of subsequent inputs allow computation of the future states of the system. The state of a system contains all the information needed to calculate future responses without reference to the history of inputs and responses. For example, the current balance of your checking account is the state of that system. There are many ways that it could have gotten to the current value, but when you are ready to write a check, that history is irrelevant. The names of the states are often described by a set of variables aptly named state variables. For systems described by differential equations, these state variables are often the independent variables of the dynamic equations. For sequential logic circuits (computers), the state variables are the outputs of the memory elements. During the execution of a computer program, the memory map and program counter provide the system state. It is important to note that the choice of state variables for a particular system is not unique—most physical systems can be described with many different sets of state variables.

A system is typically described by a state diagram, as shown in Figure 3.1. A circle indicates the state. The name of the state is above a line drawn in the middle of the circle and the value of the output is below the line. There exists a unique output for every state, but the same output can be produced from different states. Arrows connect the circles, each representing a unique combination of inputs. Every state must have the same number of arrows leaving it. In other words, each state must be defined so that it responds in some way to every input defined for the system.

Example 3.1   A two-state system for a light with an on–off switch.

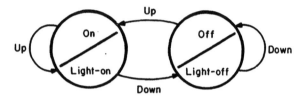

Figure 3.1 Two-state model of a light system.

State = On, Off
Input = Up, Down
Output = LightOn, LightOff

This means the state can be On or Off, the input can be Up or Down, and the output can be LightOn or LightOff. This is easy. The state and the output depend directly on the input. Now we consider a slightly more complicated system having two switches and one light.

Example 3.2   A three-state system for a light with an on–off switch on the wall and a turn-type switch on the lamp.

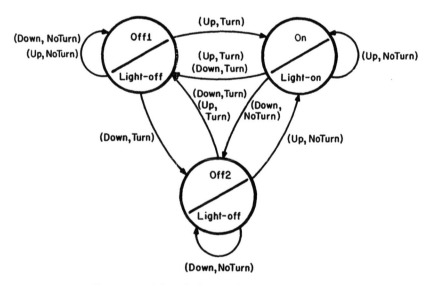

Figure 3.2 Model of a three-state light system.

        State = Off1, On, Off2
    Input s1 = Up, Down
    Input s2 = Turn, NoTurn
    Output = LightOn, LightOff

In this example, more states are needed to keep track of the switch positions. If s1 is up, the light is on or off, depending on s2. It is the change in the position of the turn switch, s2, that turns the light on or off, not simply the fact that s1 is up or down.

There are two common types of state machines: level output (Moore) machines, in which the outputs are associated with the states, and pulse output (Mealy) machines, in which the outputs are associated with the transitions between the states. The similarities and differences between these two types of machines are presented in most books on digital logic design. In this book, only level output machines are discussed.

3.2 *System design set theory*

Specifying simple system designs with drawings is convenient, but it is not practical for large systems. For this reason, a notation based on set theory will be used.

A collection of objects forms a set, the members of which are enclosed within braces. For example, the set of inputs for Example 3.1 can be expressed as {Up,Down}. The set of even positive integers, though not finite, can be expressed as {2,4,6,...}. Individual members of a set, called elements, are separated by commas. An element of a set can be a collection, such as the coordinates in the following set: {(1,1),(2,2)}. A set can be defined by enumeration, by inclusion, or by exclusion. The above sets have been defined by enumerating all elements in that set. A set can be defined by what it is not, for example, a set that does not contain negative numbers.

The Cartesian product of sets is an abbreviated way of writing all possible combinations of elements using one element from each set. The Cartesian product of Set A with Set B is expressed as A × B. If A is {a1,a2,a3} and B is {b1,b2}, then

    A × B = {(a1,b1),(a1,b2),(a2,b1),
             (a2,b2),(a3,b1),(a3,b2)}

We now define a system design notation. The form for a system model is:

    Z = (SZ, IZ, OZ, NZ, RZ)

where Z is the name of the system or system model, and SZ, IZ, OZ, NZ, and RZ are the artifacts of Z. The set SZ contains all possible states of the system.

The input set, I Z, contains all possible inputs to the system, and the output set, O Z, contains all possible system outputs. The next state function, N Z, provides a mapping of a given state from S Z with an input from I Z to form another state of S Z. Finally, R Z is the readout function, which provides a mapping of states from S Z to outputs from O Z.

Example 3.3   Redefine Example 3.1 using the system design notation.

```
Z = (SZ, IZ, OZ, NZ, RZ)
```

where

```
SZ = {On, Off},
IZ = {Up, Down},
OZ = {LightOn, LightOff},
NZ = {((On,Up), On), ((On,Down), Off),
       ((Off,Up), On), ((Off,Down), Off),},
RZ = {(On,LightOn), (Off,LightOff)}
```

The definitions of sets S Z, I Z, and O Z in this example are clear, but those for N Z and R Z are not so obvious. The first element of N Z says: If the system is in the O n state when the input U p is applied, then the next state of the system will be O n. The second element says: If the system is in the O n state when the input D o w n is applied, then the next state of the system will be O f f. The readout function, R Z, says: If the system is in the O n state, the output will be L i g h t O n; but if the system is in the O f f state, the output will be L i g h t O f f.

Example 3.4   Redefine Example 3.2 using the system design notation.

```
Z = (SZ, IZ, OZ, NZ, RZ)
```

where

```
SZ = {On, Off1, Off2},
IZ = I1 × I2,
      I1 = {Up, Down},
      I2 = {Turn, NoTurn},
OZ = {LightOn, LightOff},
NZ = {((Off1,(Up,Turn)),On),
       ((Off1,(Down,Turn)),Off2),
       ((Off1,(Up,NoTurn)),Off1),
       ((Off1,(Down,NoTurn)),Off1),
       ((On,(Up,Turn)),Off1),
       ((On,(Down,Turn)),Off1),
```

```
        ((On,(Up,NoTurn)),On),
         ((On,(Down,NoTurn)),Off2),
         ((Off2,(Up,Turn)),Off1),
         ((Off2,(Down,Turn)),Off1),
         ((Off2,(Up,NoTurn)),On),
         ((Off2,(Down,NoTurn)),Off2)},
  RZ = {(On,LightOn), (Off1,LightOff),
         (Off2,LightOff)}
```

Most systems not only have many inputs, but also have many *kinds* of inputs. For example, in a car wash system the cars coming in represent one kind of input; electrical power and water represent two more. Each different kind of input may be designated as an "input port." In considering the nature of system inputs, there is sometimes a question as to whether a model should have one input port with multiple input values or multiple input ports with fewer input values. Inputs that may occur simultaneously must be assigned to different input ports. Inputs that have different values and cannot occur simultaneously can be in the allowed set for one port. *An input port has a set of allowed values each of which is possible individually, but no two can occur simultaneously.* In our car wash example, the allowed values for the input port cars might be the license plate numbers of the cars. Different input ports could have the same set of allowed values. For example, in a bank with four tellers on duty, each of the tellers is an input port, and the allowed input values for each port would be {money, checks, withdrawal slips, etc.}. In Example 3.4, the two input ports correspond to switches s1 and s2.

Outputs have the same restrictions: *An output port has a set of allowed values each of which is possible individually, but no two can occur simultaneously.* If two outputs can occur simultaneously, then they must be values for two different output ports.

The statements in the examples above are all that is needed to describe the function of the models. However, to operate the model, a starting state and series of inputs are needed. For example, the light in Example 3.4 may be on, but the current status of the two switches is needed to determine what will happen when one of them changes.

3.3 Input and output trajectories

A trajectory is a set of data obtained over time, the time unit depending on the model. An example of an input trajectory (or an element of ITZ, where ITZ means the input trajectory set) is the customer arrivals at a bank teller's station. This input trajectory can be described by the following table:

TZ	IZ
0	1
1	0
2	3
3	2

This could be interpreted as: At opening time one person was in line. Assuming a time scale of hours, the next time increment represents what happens in the first hour; in this case, no one else arrived. Between the first and second hour, three people arrived, and between the second and third, two more arrived. This may be written $ITZ = \{(0,1),(1,0),(2,3),(3,2)\}$. The output trajectories (OTZ) are treated similarly. Trajectories are useful in that they are a way of describing data coming into the system model. Test trajectories are created to determine if given input trajectories generate specific output trajectories or a set of possible output trajectories.

Example 3.5   For Example 3.1, the following input trajectory f is provided:
$f = \{(0,Up),(1,Up),(2,Down),(3,Up),(4,Down),(5,Down)\}$.
Assume the system starts in the Off state. Provide the output trajectory.

```
OTZ(f,Off) = {(0,LightOff), (1,LightOn),
              (2,LightOn), (3,LightOff),
              (4,LightOn), (5,LightOff),
              (6,LightOff)}.
```

The notation $OTZ(f,Off)$ represents the output trajectory generated using the input trajectory f and the starting state Off. The new output takes effect on the next time interval. This is a "causal" system, since the inputs caused the outputs to occur.

3.4 System components

Large systems consist of many parts that, in many cases, function as systems in themselves—that is, they receive inputs, process them based on their current state, and produce outputs. A system that functions within another system is called a component.

Many familiar systems consist of components. One example is a stereo system, which consists of several systems that work together: the receiver, speakers, turntable, cassette player, etc. Another common system is an automobile, which has an engine, brakes, air conditioning, suspension, etc. All of these components are selected and put together in such a way that they form a larger system, yet each is a system in its own right.

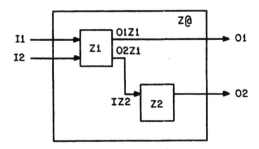

Figure 3.3 A system formed by interconnecting two components.

Components are put together using System Coupling Recipes (SCR). The outputs of one system become the inputs of another system by connecting the output port of one component to the input port of another. If feedback is needed, the output of the system is entered back into the system through another input port. For the resultant of the SCR to be a system, it must accept inputs and send outputs external to itself.

The resultant of a coupling is defined to be Z@. A system coupling recipe consists of two parts in the equation Z@ = SCR(VSCR, CSCR), where VSCR is a vector listing the systems to be connected and CSCR is the connections that must be made.

To illustrate, we define a system with two components. The first component, $Z1$, is a temperature-setting control with the switch positions Off, Low, or High. The system accepts another input for air temperature. It will output a temperature setting associated with the inputs. The second system component, $Z2$, is a heater that will accept an integer temperature input between 60 and 120 °F and produce heat at that setting, but only if the setting is above 80 °F. We couple this system together so that the second output port of $Z1$ is connected to the input port of $Z2$, as shown in Figure 3.3.

```
Z1 = {SZ1,IZ1,OZ1,NZ1,RZ1}
```

where

```
SZ1 = {Off,Low,Hi},
IZ1 = I1Z1 × I2Z1,
        I1Z1 = {Off,Low,High},
        I2Z1 = IJS[60-120],
OZ1 = O1Z1 × O2Z1,
        O1Z1 = {Off,On},
        O2Z1 = IJS[60-120],
NZ1 = {((Off,(Off,IJS[60-120]),Off),
        (Off,(Low,IJS[60-79]),Low),
        (Off,(Low,IJS[80-99]),Off),
        (Off,(Low,IJS[100-120]),Off),
```

```
            (Off,(High,IJS[60-79]),Hi),
            (Off,(High,IJS[80-99]),Hi),
            (Off,(High,IJS[100-120]),Off),
            (Low,(Off,IJS[60-120]),Off),
            (Low,(Low,IJS[60-79]),Low),
            (Low,(Low,IJS[80-99]),Low),
            (Low,(Low,IJS[100-120]),Off),
            (Low,(High,IJS[60-120]),Hi),
            (Hi,(Off,IJS[60-120]),Off),
            (Hi,(Low,IJS[60-79]),Low),
            (Hi,(Low,IJS[80-99]),Low),
            (Hi,(Low,IJS[100-120]),Off),
            (Hi,(High,IJS[60-120]),Hi)},
    RZ1 = {(Off,(Off,0)), (Low,(On,90)),
            (Hi,(On,110))}.
```

and

```
    Z2 = {SZ2,IZ2,OZ2,NZ2,RZ2}
```

where

```
    SZ2 = IJS[60-120],
    IZ2 = IJS[60-120],
    OZ2 = IJS[60-120],
    NZ2 = {((x,y),x): x ∈ IZ2, y ∈ SZ2},
    RZ2 = {(x,x): x ∈ SZ2}.
```

The notation $I1Z1$ is used for the first input to component $Z1$. The second input port to $Z1$ is $I2Z1$, which is an integer (denoted by the function IJS) between 60 and 120.

For $Z2$, we have a simple next state function in which the next state is based entirely on the input; that is, if the input is 100, then no matter what the current state is, the next state is 100. This is described by the line $NZ2 = \{((x,y),x): x \in IZ2, y \in SZ2\}$. In this expression, the input to the function is x, and it is currently in state y. The symbol $\in$ indicates that the input x is an element of the input set $IZ2$, and the state y is an element of the state set $SZ2$. The next state is x. This only works if the input set is a subset of the states.

In $Z2$ we have state readout for $RZ2$; that is, the output matches the state for all states. This is described by the set notation $\{(x,x): x \in SZ2\}$, which says that the output of the function is the state.

The two systems are connected or coupled as described by the following system coupling recipe:

```
    SCR = ((Z1,Z2), {(O2Z1,IZ2)}).
```

This expression is interpreted as: $Z1$ is coupled with $Z2$ by connecting the second output port of $Z1$, which is $O2Z1$, to the input port of $Z2$, which is $IZ2$.

A common problem that occurs when connecting components is mismatched input ports (or output ports). This happens when communications between components are not modeled correctly, resulting in the data expected by one component not being what is received. For example, let us assume that one system component, Z6, has an output port defined to send the data set $\{0, 2, 4\}$ out and another component, Z7, has an input port designed to accept the data set $\{0,1,2,3,4\}$. In this case, no errors are created, but Z7 only uses a subset of its capability (see system modes below). However, if the situation is changed so that the output sends $\{0,1,2,3,4\}$ and the input receives only $\{0,2,4\}$, then a mismatch does occur and the system may enter an unknown state. In software this is often called "a bug," but in reality a subroutine (the software equivalent of a component) has received undefined inputs that cause an error.

During the design of a system, the components should be chosen in such a way that they have about equal complexity because it is hard to manage the design of a system when one component is trivial and another is extremely complex. For example, referring for the moment back to Exhibit 2.4, the fine guidance control system of the Hubble telescope was on the cutting edge of technology. In fact, for a while it looked like it could not be built to meet the specifications. However, after substantial time and budget overruns, a successful system was built. By comparison, grinding the mirror was a trivial task, but the primary mirror was ground in the wrong shape and there was no full system test to disclose this mistake. Why not? One of the main reasons was that the fine guidance control system had so drastically overrun the budget that corners were cut on all the other components.

3.5 System modes

Sometimes the required system input and output behavior can be found in a part of a bigger system. When such a part is a system in itself we call it a system mode. Often it is less expensive to buy the big system and use only the part of it that is needed to satisfy the input and output behavior. For example, for Christmas one year, Dr. Bahill's sons, Alex and Zach, said they wanted radios—AM for Alex and FM for Zach. It is very hard to find an AM only or an FM only radio, since most radios have both. Dr. Bahill decided to buy two AM–FM radios and only use the wanted system mode on each. That is, he put Alex's switch in the AM position and Zach's switch in the FM position, and they were as happy as brothers can be.

Note that system modes are different from components. The components of a radio are an FM antenna, an AM antenna, an FM tuner, an AM tuner, an amplifier, a speaker, a battery, and a case. The FM system mode uses a subset of these components, namely, the FM antenna, the FM tuner, the amplifier, the speaker, the battery, and the case.

3.6 System homomorphisms

The results of many systems are the same, or they can be made to look the same with a transformation. A homomorphism is a function that translates three sets from one system into another system: the set of states (SZ), the set of inputs (IZ), and the set of outputs (OZ). An important use of a homomorphism is for translating the results of a model into a real world application.

Homomorphisms to change $Z1$ into $Z2$ use the following syntax:

```
HS = {(SZ1,SZ2)}, where HS is a function mapping SZ1 onto SZ2.
HI = {(IZ1,IZ2)}, where HI is a function mapping IZ1 onto IZ2.
HO = {(OZ1,OZ2)}, where HO is a function mapping OZ1 onto OZ2.
```

Homomorphisms are defined as many to one; that is, the mapping from $Z1$ to $Z2$ is not necessarily reversible from $Z2$ to $Z1$. For example, squaring the value of the inputs, (–1, 2, –3) becomes (1, 4, 9); that is, –1 maps to 1, 2 maps to 4, and –3 maps to 9. But the inverse mapping, using the square root function, yields (1, 2, 3), which is a different input set.

An isomorphic mapping is a one-to-one mapping, in which the sets are simply renamed from one system to another.

Example 3.6   Create another input set for Example 3.4 that is isomorphic to the original.

```
IZ = I1 × I2                          The original set.
     I1 = {Up, Down},
     I2 = {Turn, NoTurn}.

IZ1 = I1 × I2                         The isomorphic set.
      I1 = {0, 1},
      I2 = {0, 1}. ·
```

Example 3.7   Use a homomorphic mapping to translate our model in Example 3.3 into the electrical wiring system of Figure 3.4.

```
Z = (SZ, IZ, OZ, NZ, RZ)
```

where

```
SZ = {On, Off},
IZ = {Up, Down},
```

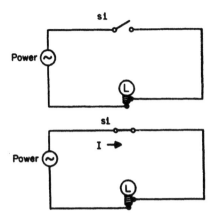

Figure 3.4 Schematic diagram of a light system: (*top*) switch s1 is open, no current flows, and the light is off; (*bottom*) switch s1 is closed, current I flows, and the light is on.

```
OZ = {LightOn, LightOff},
NZ = {((On,Up), On), ((On,Down), Off),
      ((Off,Up), On), ((Off,Down), Off),},
RZ = {(On,LightOn), (Off,LightOff)}.
HS = {(On, I flowing), (Off, I stopped)},
HI = {(Up,s1 closed), (Down,s2 open)},
HO = {(LightOn, L on), (LightOff, L off)}.
```

Below is a list of all the acronyms used in these models.

CSCR	Connections for a system coupling recipe
HI	Homomorphic input mapping
HO	Homomorphic output mapping
HS	Homomorphic state mapping
ITZ	Set of system inputs over time
IZ	Set of system inputs
NZ	Next state function
OTZ	Set of system outputs over time
OZ	Set of system outputs
RZ	Readout function
SCR	System coupling recipe
SZ	Set of system states
TZ	Time trajectory
VSCR	Vector of systems to couple

3.7   System modeling

Modeling is an important facet of engineering. Models are simplified representations of real world systems. For this reason, models are also an important part of everyday life. If you wanted to drive from Tucson, Arizona to Bethlehem, Pennsylvania, you would use a model of the highway system called a road map. The road map allows you to study and understand the highway system of the United States without having to drive on every road. If you wanted a model of the life of southern aristocrats at the time of the Civil War, you might study the book *Gone with the Wind*. Highway engineers create models to see how traffic-light synchronization or lane obstructions affect traffic flow. Ohm's Law ($E = IR$) is a good model for a resistor. The equation $F = ma$ is a simple model for the movement of a baseball. Models allow us to apply mathematical tools to real world systems. We use models to understand things that are big, complicated, expensive, or far away in space or time.

Figure 3.5 is an overview of people studying systems. On the extreme left, people are doing experiments on real world systems. Baseball players often fit into this category.

Over the years, baseball players have experimented a great deal with the baseball bat. Most of this experimentation was illegal, because the rules (for professional players) say that the bat must be made from one solid piece of wood. George Sisler, who played first base for the St. Louis Browns in the 1920s, pounded Victrola phonograph needles into his bat barrel to make it heavier; and in the 1950s, Ted Kluszewski of the Cincinnati Reds hammered in big nails. To make the bat lighter, many players have drilled a hole in the end of the bat and filled it with cork. Detroit's Norm Cash admits to using a corked bat in 1961, when he won the American League batting title by hitting .361. However, the corked bat may have had little to do with his success, because he presumably used a corked bat the next year when he slumped to .243. Some players have been caught using doctored bats. In 1974, the bat of Graig Nettles of the Yankees shattered when it hit the ball and out bounced six super balls. In 1987, Houston's Billy Hatcher hit the ball and his bat split open, spraying cork all over the infield. These are all examples of individuals experimenting with no models to guide them.

To lessen the wasted time and reduced performance entailed in such experiments with altered bats, we made mathematical models of individual humans and coupled these models to physics equations to predict the ideal bat weight for each individual (Bahill and Karnavas, 1991). This modeling process is shown in the left half of Figure 3.5.

We next address the box on the right side of Figure 3.5 labeled Computer simulation of the model. Our model was composed of mathematical equations that had to be solved on a computer. If everything goes right, the digital computer simulation should produce the same results as the mathematical equations. But care must be taken to ensure that this is true. Topics for concern include (1) the accuracy of the computer code; (2) numerical factors (for details

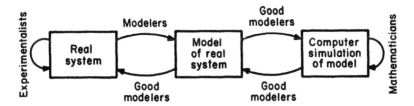

Figure 3.5 Relationships between simulations, models, and the real world.

see Yakowitz and Szidarovszky, 1989) such as the integration step size, truncation errors, and the integration technique (e.g., quadrature, Adams, Runge–Kutta) ; (3) implementation considerations, such as using a commercial simulation package that is much bigger than the model (e.g., using a calculator to add single digit numbers so that in some situations the unused routines could cause problems by overwriting areas of memory or forcing pointers out of bounds); and (4) the possibility that the hardware is defective (How often do you run the diagnostics on your personal computer?). In our study we carefully assessed each of these to see how they would affect our predictions about the real world.

Finally, at the extreme right of Figure 3.5 we find pure mathematicians working in the computer world with no regard to the real world. Early studies of fractals were performed by pure mathematicians.

For a further discussion of the philosophy and practice of modeling, see Clements (1989), Bahill (1981), or Jeffreys and Berger (1992).

Models are especially used in system design and analysis. Conceptual designs will change as a model is built. The problems in communications between components, the matching of inputs and outputs, and the unambiguous representation of states will all become clear as the model is perfected. The analysis of the model is often cheaper and easier than using a prototype, but the data collected from the analysis is only as good as the original model and should be treated as such.

The modeling language introduced in this chapter is actually much more elaborate than the usage in the examples above (see Wymore, in press), but we have presented only enough to be able to understand the modeling used in the Pinewood and SIERRA case studies. There is no agreed upon modeling language. State diagrams are common, but become difficult to use when the number of states, inputs, or outputs becomes large. Other modeling languages are briefly described in Chapter 7.

One of the major aspects of dynamic modeling that is handled poorly by state diagrams is the representation of time. Time phase diagrams are very common in engineering applications. "What happens when?" is critical knowledge for complex systems. The question may be answered by creating a model using state diagrams and then varying the input trajectories, but a visual presentation is lacking. The actual diagrams needed to represent time-dependent systems would be enormous, since every time-varying state must be shown separately. The set theoretic notation presented above does not have this drawback, since the states can be infinite if need be.

Another time-dependent modeling task that is not handled well by state diagrams is the process flow diagrams used by industrial engineers and management science practitioners. These diagrams show the sequential flow through an area. A state diagram can be used to model sequential processes, but again it becomes large and the processes are difficult to represent.

The most common tool for modeling dynamic systems is state–space notation, which is the topic of Szidarovsky and Bahill (1992). This modeling technique is often used in conjunction with the set theoretic notation introduced in this chapter. For example, for a computer that controls a squirrel-cage induction motor, the set theoretic notation is more convenient for the computer, whereas state–space notation is more convenient for the induction motor.

Although state representation is a useful modeling tool, particularly for initial system design, it is not the only tool for modeling. In industry, system designers will generally use whatever tools are available. If a different modeling tool—such as a simulation language—exists, then it will be used. The design process used to create a state representation will be valid, even if the modeling tool is different. It is often helpful in generating functional diagrams to use the state modeling described above to ensure that all inputs and outputs are properly represented and that the connections are made; then translate the diagrams into an available modeling language. The next chapter discusses in detail how to model the requirements of a system and how to use the model to select the best system design concept.

Problems

1. **Set theory.** These problems are intended to help familiarize you with our set theoretic notation. Let

```
A = IJS[1-9],
/*The notation IJS[1-9] in these homework
   problems means the integers 1 to 9.*/
B = {x: x ∈ A; x is odd},
/*This means B is the set of all x where x is an
   element of the set A and x is odd.*/
C = {x: x ∈ A; x is even},
D = IJS[3-5],
E = {x: x ∈ D; x is odd}.
```

These sets can also be defined by enumeration as follows:

A = {1, 2, 3, 4, 5, 6, 7, 8, 9},
B = {1, 3, 5, 7, 9},
C = {2, 4, 6, 8},
D = {3, 4, 5},
E = {3, 5}.

Find:

(i) B ∪ C. This is called B union C.

(ii) B × D. This is the Cartesian product of sets B and D.

(iii) E ∩ C. This is called the intersection of E and C.

(iv) A – B. This is often expressed as A ∩ $\overline{B}$.

(v) Deduce the possible sets X for which X ∩ B = { }.

Define F by enumeration, where F = {1, 2}^3; that is F = {1, 2} × {1, 2} × {1, 2}.

2. **Automobile ignition circuit.** Derive a state diagram for a system to enable the ignition circuit of an automobile only if the driver first sits down in the seat and then fastens the seat belt. Label the states with meaningful names. State all the assumptions you make. Call this system Z2, and specify

Z2 = (SZ2, IZ2, OZ2, NZ2, RZ2).

A general technique for creating state diagrams is (1) define a start state, (2) draw m^n arrows leaving this state, where n is the number of input ports and m is the number of values each port can have (assuming each port can accept the same number of values), (3) add states as needed so that all arrows terminate on a state, (4) repeat steps 2 and 3 until all possible transitions have been accounted for, (5) remove the start state, if necessary, and (6) minimize, if desired.

---

† Construct a state table and derive state equations for this system. Use RS flip-flops. You will probably have an unused state. What happens to your system if it inadvertently gets into this unused state, either when power is first applied or due to noise?

3. **Odd parity detector.** Assume that computer words are transmitted in binary form over a wire. Draw a state diagram for a system whose output will be 1 if, and only if, the present and previous bits showed odd parity; that is, the output is 1 if the previous two bits were 01 or 10, and it stays at 1 until

---

† This section requires technical material not presented in this text.

the next bit is received. Conversely, the output is 0 if the bits were 00 or 11. Call this system $Z3$, and specify $Z3 = (SZ3, IZ3, OZ3, NZ3, RZ3)$. Describe a test for your model that would assure its correctness.

Defining inputs and states is usually the most difficult task in designing systems. Inputs to state machines should be simple. Typical inputs are switches that can be open or closed. Inputs should not be described as having memory. For example, if you are trying to detect the presence of two consecutive 1's or two consecutive 0's in a communications line, you should not describe the inputs with the phrases "present bit" and "previous bit." Furthermore, inputs should not be described as intelligent machines. For example, if you were to design a system for a space shuttle launch to determine if all pre-launch operations (e.g., load crew, close door, load oxygen, load hydrogen, release clamps) have been done in the correct order, then your input should not be:

  $x = 1$ means all operations were done in correct order and
  $x = 0$ means operations were not done in correct order.

Names of states *should* reflect concern for the sequence of past events. They should not sound like combinations of present inputs. For example, for the automobile seat belt problem, "Seat Occupied and Seat Belt Fastened" is an incorrect choice for a state name, but "Seat Occupied *then* Seat Belt Fastened" might be appropriate. In general, if a state has the same name as an input, then you have probably made a mistake. It is acceptable for a state name to be the same as that of an output, in which case you have merely used the state readout.

For most of our problems you may, if you wish, ignore start-up states and assume that the system is operating in steady-state.

---

4. **System permutations.** Draw a state diagram for the system $Z4$ described below. If you were to change only the function NZ, how many different systems could you make?

    $Z4 = (SZ4, IZ4, OZ4, NZ4, RZ4)$

where

    $SZ4 = \{A, B\}$,
    $IZ4 = \{Yes, No\}$,
    $OZ4 = \{On, Off\}$,
    $NZ4 = \{((A, No), A), ((A, Yes), B),$
            $((B, No), B), ((B, Yes), A)\}$,
    $RZ4 = \{(A, Off), (B, On)\}$.

5. **Input port structure.** Draw a state diagram and deduce the input port structure for the following system, $Z5$.

$$Z5 = \{SZ5, IZ5, OZ5, NZ5, RZ5\}$$

where

```
SZ5 = {Red, White},
IZ5 = ?
OZ5 = {On, Off},
NZ5 = {((Red, (1, 3)), Red),
        ((Red, (1, 4)), White),
        ((Red, (2, 3)), Red),
        ((Red, (2, 4)), White),
        ((White, (1, 3)), White),
        ((White, (1, 4)), Red),
        ((White, (2, 3)), White),
        ((White, (2, 4)), Red)},
RZ5 = {(Red, On), (White, Off)}.
```

6. **Three-way switches.** Create a model of a household three-way switch system in which the light is controlled by switches at each end of a hallway. Use state readout and let the state and the output be the condition of the bulb whether On or Off. Let the inputs be NC if there has been no change in switch position, and CH if there has been a change in switch position. Call this system $Z6$; specify $Z6 = (SZ6, IZ6, OZ6, NZ6, RZ6)$.

7. **Invalid BCD digit detector.** In the Binary Coded Decimal (BCD) representation, each decimal digit is represented with four binary bits. The numbers 0 to 9 are represented by valid four-bit BCD digits, and the numbers 10 to 15 constitute invalid BCD digits. Assume the four bits are being transmitted on a serial line. Draw a state diagram for a single-input, single-output system that determines whether an invalid BCD digit has been received. If, at the end of the BCD word (the four bits), your system determines that the BCD digit is invalid, then its output should be 1 and remain 1 until another bit is received. The output should be 0 at all other times. Least-significant bits of the words appear first in time. You may ignore start-up transients; design your system for steady-state operation. You need not minimize your system. It will be easier if you treat the input as a sequence of words and not as a string of bits. That is, test bits 1, 2, 3, and 4, then bits 5, 6, 7, and 8; do not test bits 1, 2, 3, and 4, then bits 2, 3, 4, and 5. Call this system $Z7$; specify $Z7 = (SZ7, IZ7, OZ7, NZ7, RZ7)$. If you wanted to test your model by applying all possible input trajectories and all reasonable initial states, how many test trajectories would you need and how long would they have to be?

8. **Input ports.** How many input ports should there be in a model for

(i) a laboratory door combination lock? Typically these locks have five buttons that can be pressed alone or in combination and a door knob.

(ii) a combination lock for a school locker? The lock requires you to first turn the dial two complete turns clockwise and stop on the first combination number, turn it one complete turn counterclockwise and stop on the second number, and finally turn it clockwise and stop on the final number.

(iii) a pushbutton telephone in normal operation?

(iv) a keyboard connected to an IBM-compatible personal computer?

(v)   a human being?

You may give short explanations for your answers if you wish.

9. **Candy machine.** In the SIE geedunk machine, candy bars cost 15¢ (OK, it's an old machine). The machine sorts coins into three categories: nickels, dimes, and others. Draw a state diagram for a system that will drop a candy bar if the customer deposits either three nickels or a nickel and a dime. If the customer deposits foreign coins or illegal combinations (such as two dimes), all deposited coins are returned. Describe your inputs, outputs, and states. Label your states with meaningful names. State all assumptions that you make. Call this system $Z9$; specify $Z9 = (SZ9, IZ9, OZ9, NZ9, RZ9)$.

10. **Telephone system.** In many parts of the United States it is not necessary to dial an initial 1 to tell the system you are placing a long distance phone call. You merely dial the area code and the system figures it out. How does it do this? All area codes have a 1 or a 0 as the second digit; no phone numbers do (except for 911—emergency). Design a system that will turn on a light after the third number if it is an area code. The light should stay on until the customer hangs up. Describe your inputs, outputs, and states. Label your states with meaningful names. State all assumptions that you make. Call this system $Z10$; specify $Z10 = (SZ10, IZ10, OZ10, NZ10, RZ10)$.

11. **Spelling checker.** Design a system to detect spelling errors. Or more simply, implement the spelling rule "i before e except after c." If a word violates this rule, your system should stop processing words and turn on an error light. When the system operator acknowledges the mistake and turns off the error light, your system should resume processing words. For example, the words "piece" and "receive" are correct so your system should continue processing words. However, "yeild" and "weird" violate this rule, so your system should stop and wait for operator action. You may assume that bizarre sequences such as "ceie" will go undetected and that Professor Lucien Duckstein will not use it. Describe your inputs, outputs, and states. Label your states with meaningful names. Assume that the system starts in a reset state. State all assumptions that you make. Call this system $Z11$; specify $Z11 = (SZ11, IZ11, OZ11, NZ11, RZ11)$.

How can you test your system?

12. **Combination lock.** A new combination lock has recently been installed on the door of our laboratory. It has five buttons that can be pressed individually or in combination and a door knob that can be turned clockwise only. Presume that the correct combination is to push buttons 4 and 2 simultaneously and then push button 3. Turning the door knob clockwise opens the door and resets the lock. Make a model of this system. Describe your inputs, outputs, and states. Draw a state diagram. Label your states with meaningful names. State all assumptions that you make. Call this system Z12; specify Z12 = (SZ12, IZ12, OZ12, NZ12, RZ12).

Now presume that you are not allowed to push two or more buttons at the same time and that the combination is changed to 4, then 3. Describe the necessary changes in your system. You need not specify the new system, just the changes. Words are sufficient.

13. **Minimization.** We have designed a system, Z13 described below, to detect two consecutive heads in a coin flipping contest. We do not think the system is minimal. Minimize it using any technique you wish, but explain your work.

> *Definition:* States are equivalent if they give exactly the same output for each member of a set of inputs and send the system either to the same state or to an equivalent state. Equivalent states can be combined into one state.

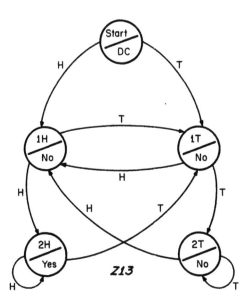

```
Z13 = (SZ13, IZ13, OZ13, NZ13, RZ13)
```

where

```
SZ13 = {1H, 2H, 1T, 2T},
IZ13 = {H, T, SOE},                                          !

/*H represents heads, T represents tails, and
  SOE represents stands on edge*/

OZ13 = {Yes, No},
NZ13 = {((1H, H), 2H), ((1H, T), 1T),
        ((1H, SOE), 1H), ((2H, H), 2H),
        ((2H, T), 1T), ((2H, SOE), 2H),
        ((1T, H), 1H), ((1T, T), 2T),
        ((1T, SOE), 1T), ((2T, H), 1H),
        ((2T, T), 2T), ((2T, SOE), 2T)},
RZ13 = {(1H, No), (2H, Yes), (1T, No), (2T, No)}.
```

The state diagram in the figure is not the same as the set theoretic description. The start-state in the figure is not a part of the steady-state system described above. Often it is helpful to begin your design with a start-state and erase it later. Also, the possibility of the coin standing on edge is not shown in the figure.

14. **State variables.** How many state variables should there be in a model for

(i)   the watch on your wrist?

(ii) a simple calendar?

(iii) the odd parity detector of Problem 3?

(iv) the human body, as assessed by a nurse in a simple physical exam?

(v)   your TV set?

(vi) a baseball game, so that you could resume a rained out game?

† (vii) a computer, immediately before executing a jump to subroutine instruction?

Describe the state variables you have envisioned.

15. **Gasoline pumps.** Some gasoline stations have a computer system in each pump that asks if you want to pay with cash or a credit card. If you say cash, it will charge you $1.00 per gallon; if you say credit card, it will charge $1.04 per gallon. Then it asks if you want to buy $5, $10, $15 or fill up the tank. (To make this problem simpler, let us ignore the possibility of filling up the

---

† This section requires technical material not presented in this text.

tank.) When the running total you owe equals the amount you chose, the computer turns off the pump. Design such a system. Describe your inputs, outputs, and states. Draw a state diagram. Label your states with meaningful names. State all assumptions that you make. Call this system Z15; specify Z15 = (SZ15, IZ15, OZ15, NZ15, RZ15).

16. † **Homomorphic images.** A single-input, single-output system, Z161, is described below. It detects invalid Binary Coded Decimal (BCD) words. At the end of each word, if the four bits constitute an invalid BCD digit (the binary equivalents of 10 to 15) the output is a 1. Otherwise, the output is a 0. Least significant bits of the words appear first in time. (Yes, Z161 is the same as Z7 of Problem 7.) Another system, Z162, also described below, is a homomorphic image of Z161. Find the mapping between these two systems. *Hint:* it might help to minimize Z161.

Definition: Two models for the same system are called homomorphic images. Sometimes the mapping between the two models is just a renaming of the inputs, outputs, and states. Sometimes one of the models is a simplification or an elaboration of the other.

Z161 = (SZ161, IZ161, OZ161, NZ161, RZ161)

where

SZ161 = {a, b, c, d, e, f, g, h, i, j, k, l, m,
         n, p, q},
IZ161 = {0, 1},
OZ161 = {0, 1},
NZ161 = {((a, 0), b), ((a, 1), c), ((b, 0), d),
         ((b, 1), e), ((c, 0), f), ((c, 1), g),
         ((d, 0), h), ((d, 1), i), ((e, 0), j),
         ((e, 1), k), ((f, 0), l), ((f, 1), m),
         ((g, 0), n), ((g, 1), p), ((h, 0), a),
         ((h, 1), a), ((i, 0), a), ((i, 1), q),
         ((j, 0), a), ((j, 1), q), ((k, 0), a),
         ((k, 1), q), ((l, 0), a), ((l, 1), a),
         ((m, 0), a), ((m, 1), q), ((n, 0), a),
         ((n, 1), q), ((p, 0), a), ((p, 1), q),
         ((q, 0), b), ((q, 1), c)},
RZ161 = {(a, 0), (b, 0), (c, 0), (d, 0), (e, 0),
         (f, 0), (g, 0), (h, 0), (i, 0), (j, 0),

---

† This problem may require tools not presented in the text.

$(k, 0)$, $(l, 0)$, $(m, 0)$, $(n, 0)$, $(p, 0)$, $(q, 1)\}$.

and

```
Z162 = (SZ162, IZ162, OZ162, NZ162, RZ162)
```

where

```
SZ162 = {A, B, C, D, E, F, G},
IZ162 = {No, Yes},
OZ162 = {Off, On},
NZ162 = {((A, No), F), ((A, Yes), F),
         ((B, No), F), ((B, Yes), F),
         ((C, No), A), ((C, Yes), B),
         ((D, No), G), ((D, Yes), C),
         ((E, No), C), ((E, Yes), C),
         ((F, No), D), ((F, Yes), E),
         ((G, No), A), ((G, Yes), A)},
RZ162 = {(A, Off), (B, On), (C, Off), (D, Off),
         (E, Off), (F, Off), (G, Off)}.
```

17. **Campus phone system.** We need only dial five numbers to get someone on campus; all main campus numbers begin with a 1 and all medical school numbers begin with a 6. To get an outside line we dial 9. To call long distance we dial 82 unless it is inside Arizona, in which case we dial 81 instead. Of course, 0 calls the operator. Design a system that will monitor a telephone and turn on an error light if a mistake is made. The light should stay on until the person hangs up. You may ignore the * and # buttons. Describe your inputs, outputs, and states. Draw a state diagram. Label your states with meaningful names. State all assumptions that you make. Call this system Z17; specify Z17 = (SZ17, IZ17, OZ17, NZ17, RZ17).

18. †**Homomorphisms.** Distinguish between a copy, an isomorphic image, and a homomorphic image.

19. **System coupling recipes.** For systems Z191 and Z192, draw (or specify CSCR in set notation) all system coupling recipes that yield valid systems, that is systems with at least one external input and output.

```
Z191 = {SZ191, IZ191, OZ191, NZ191, RZ191}
```

where

```
SZ191 = {Aa, Bb},
IZ191 = {1, 2} × {3, 4},
OZ191 = {4, 5},
```

---

† This problem may require tools not presented in the text.

```
NZ191 = {(((Aa, (1, 3)), Aa), ((Aa, (1, 4)), Bb),
          ((Aa, (2, 3)), Aa), ((Aa, (2, 4)), Bb),
          ((Bb, (1, 3)), Bb), ((Bb, (1, 4)), Aa),
          ((Bb, (2, 3)), Bb), ((Bb, (2, 4)), Aa)},
RZ191 = {(Aa, 4), (Bb, 5)}.
```

and

```
Z192 = {SZ192, IZ192, OZ192, NZ192, RZ192}
```

where

```
SZ192 = {Cc, Dd},
IZ192 = {1, 2} × {4, 5},
OZ192 = {1, 2},
NZ192 = {(((Cc, (1, 5)), Cc), ((Cc, (1, 4)), Dd),
          ((Cc, (2, 4)), Cc), ((Cc, (2, 5)), Dd),
          ((Dd, (1, 5)), Dd), ((Dd, (1, 4)), Cc),
          ((Dd, (2, 4)), Dd), ((Dd, (2, 5)), Cc)},
RZ192 = {(Cc, 1), (Dd, 2)}.
```

20. † **System modes-1.** Your boss wants to build systems Z202 and Z203, which are described below and shown in the figure. However, you notice that system Z201 is right over there on the shelf, and you think that Z202 and Z203 are both system modes of Z201. So, you propose to merely use Z201, controlling its initial state appropriately to get the behavior you want, whether that be of system Z202 or Z203. Design a system experiment to prove this.

```
Z201 = {SZ201, IZ201, OZ201, NZ201, RZ201}
```

where

```
SZ201 = {A, B, C, D},
IZ201 = {1, 2} × {3, 4, 5},
OZ201 = {Yes, No, Go, Stop},
NZ201 = {((A, (1, 3)), A), ((A, (1, 4)), B),
          ((A, (1, 5)), C), ((A, (2, 3)), A),
          ((A, (2, 4)), B), ((A, (2, 5)), C),
          ((B, (1, 3)), B), ((B, (1, 4)), A),
          ((B, (1, 5)), D), ((B, (2, 3)), B),
          ((B, (2, 4)), A), ((B, (2, 5)), D),
          ((C, (1, 3)), A), ((C, (1, 4)), D),
          ((C, (1, 5)), C), ((C, (2, 3)), A),
          ((C, (2, 4)), C), ((C, (2, 5)), D),
          ((D, (1, 3)), B), ((D, (1, 4)), C),
```

---

† This problem may require tools not presented in the text.

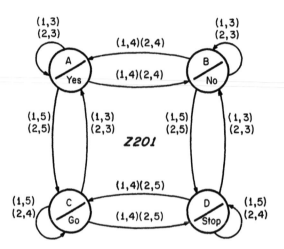

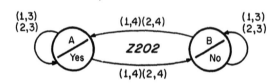

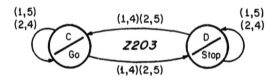

```
        ((D, (1, 5)), D), ((D, (2, 3)), B),
        ((D, (2, 4)), D), ((D, (2, 5)), C)},
RZ201 = {(A, Yes), (B, No), (C, Go), (D, Stop)}.
```

and

```
Z202 = {SZ202, IZ202, OZ202, NZ202, RZ202}
```

where

```
SZ202 = {A, B},
IZ202 = {1, 2} × {3, 4},
OZ202 = {Yes, No},
NZ202 = {((A, (1, 3)), A), ((A, (1, 4)), B),
         ((A, (2, 3)), A), ((A, (2, 4)), B),
```

```
              ((B, (1, 3)), B), ((B, (1, 4)), A),
              ((B, (2, 3)), B), ((B, (2, 4)), A)},
  RZ202 = {(A, Yes), (B, No)}.
```

. and

```
  Z203 = {SZ203, IZ203, OZ203, NZ203, RZ203}
```

where

```
  SZ203 = {C, D},
  IZ203 = {1, 2} × {4, 5},
  OZ203 = {Go, Stop},
  NZ203 = {((C, (1, 5)), C), ((C, (1, 4)), D),
           ((C, (2, 4)), C), ((C, (2, 5)), D),
           ((D, (1, 5)), D), ((D, (1, 4)), C),
           ((D, (2, 4)), D), ((D, (2, 5)), C)},
  RZ203 = {(C, Go), (D, Stop)}.
```

21. † **System modes–2.** Your boss wants to build systems $Z212$ and $Z213$, which are described below and shown in the figure. However, you notice that system $Z211$ is right over there on the shelf, and you know that $Z212$ and $Z213$ are both system modes of $Z211$ (you do not have to prove this). So, you propose to merely use $Z211$, controlling its initial state appropriately to get the behavior you want, whether that be of system $Z212$ or $Z213$. Did you do a good job? Do you deserve a bonus or do you deserve to be fired? Why?

```
  Z211 = {SZ211, IZ211, OZ211, NZ211, RZ211}
```

where

```
  SZ211 = {Red, White, Blue, Green},
  IZ211 = {1, 2} × {3, 4, 5},
  OZ211 = {On, Off, Hot, Cold},
  NZ211 = {((Red, (1, 3)), Red),
           ((Red, (1, 4)), White),
           ((Red, (1, 5)), Blue),
           ((Red, (2, 3)), Red),
           ((Red, (2, 4)), White),
           ((Red, (2, 5)), Blue),
           ((White, (1, 3)), White),
           ((White, (1, 4)), Red),
           ((White, (1, 5)), Green),
           ((White, (2, 3)), White),
```

---

† This problem may require tools not presented in the text.

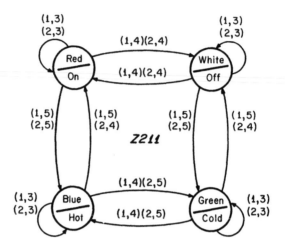

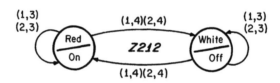

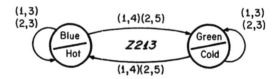

```
((White, (2, 4)) ,Red),
((White, (2, 5)), Green),
((Blue, (1, 3)), Blue),
((Blue, (1, 4)), Green),
((Blue, (1, 5)), Red),
((Blue, (2 ,3)), Blue),
((Blue, (2, 4)), Red),
((Blue, (2, 5)), Green),
((Green, (1, 3)), Green),
((Green, (1, 4)), Blue),
((Green, (1, 5)), White),
((Green, (2, 3)), Green),
((Green, (2, 4)), White),
```

```
                ((Green, (2, 5)), Blue)},
   RZ211 = {(Red, On), (White, Off), (Blue, Hot),
            (Green, Cold)}.
```

and

```
   Z212 = {SZ212, IZ212, OZ212, NZ212, RZ212}
```

where

```
   SZ212 = {Red, White},
   IZ212 = {1, 2} × {3, 4},
   OZ212 = {On, Off},
   NZ212 = {((Red, (1, 3)), Red),
            ((Red, (1, 4)), White),
            ((Red, (2, 3)), Red),
            ((Red, (2, 4)), White),
            ((White, (1, 3)), White),
            ((White, (1, 4)), Red),
            ((White, (2, 3)), White),
            ((White, (2, 4)), Red)},
   RZ212 = {(Red, On), (White, Off)}.
```

and

```
   Z213 = {SZ213, IZ213, OZ213, NZ213, RZ213}
```

where

```
   SZ213 = {Blue, Green},
   IZ213 = {(1, 3), (2, 3), (1, 4), (2, 5)},
   OZ213 = {Hot, Cold},
   NZ213 = {((Blue, (1, 3)), Blue),
            ((Blue, (1, 4)), Green),
            ((Blue, (2, 3)), Blue),
            ((Blue, (2, 5)), Green),
            ((Green, (1, 3)), Green),
            ((Green, (1, 4)), Blue),
            ((Green, (2, 3)), Green),
            ((Green, (2, 5)), Blue)},
   RZ213 = {(Blue, Hot), (Green, Cold)}.
```

22 † **Isomorphisms.** Are any of the systems $Z201, Z202, Z203, Z211,$ $Z212,$ or $Z213$ of Problems 20 and 21, as shown in the figures for those problems, isomorphic images of each other? If so, provide the relationships— that is, $HS, HI,$ and HO.

---

† This problem may require tools not presented in the text.

Isomorphic images are similar to homomorphic images, as defined in
Problem 16, with the exception that isomorphic images must have the
same number of states.

23. † **Controllability and observability.**   Draw the state diagram for the
system described below. Fill in the table with yes or no answers, depending
on whether State y is reachable from State x. Let

$$Z23 = (SZ23, IZ23, OZ23, NZ23, RZ23)$$

where

```
SZ23 = {A1, A2, A3},
I1Z23 = {6, 7},
OZ23 = {4, 5},
NZ23 = {((A1, 6), A2), ((A1, 7), A1),
         ((A2, 6), A2), ((A2, 7), A1),
         ((A3, 6), A3), ((A3, 7), A1)},
RZ23 = {(A1, 4), (A2, 5), (A3, 5)}.
```

	State y, "to"		
State x, "from"	A1	A2	A3
A1			
A2			

Simple definitions: A system is controllable if there exists an input
trajectory that will put the system into any given state. A system is
observable if you can know what the initial state was by looking at the
output trajectory. Unminimized systems are not observable.

Provide reasons for your answers to the following questions:

    (i) Is this system completely controllable to state A3?

    (ii) Is this system completely controllable from state A3?

    (iii) Is this system completely observable?

---

† This problem may require tools not presented in the text.

24. **The wolf, goat, cabbage riddle.** On the left bank of a river in the forest there is a traveler with his large dog (which is part wolf), his goat, and two dozen heads of cabbage. He wishes to reach a town on the right bank of the river with all his possessions in a small boat that has a capacity for him and only one of his charges. His task is complicated by the fact that if left alone the wolf will eat the goat and the goat will eat the cabbage. Furthermore, he does not want the cabbage sitting alone on the bank by the forest, because the mice in the forest might eat it. (Attributed to Alcuin, a friend of Charlemagne.)

This riddle can be solved as a system design problem. Let

$T = 1$ represent the traveler on left bank (by implication $\bar{T}$, or $T = 0$, represents the traveler on the right bank),
$W = 1$ represent the wolf on left bank,
$G = 1$ represent the goat on left bank, and
$C = 1$ represent the cabbage on left bank.

Find the sequence of states that will take the traveler and his possessions safely from the left bank ($TWGC = 1111$) to the right bank ($TWGC = 0000$). Explain your work.

25. † **Accommodating continuous systems.** Make a model for a transistor audio amplifier, stating $Z25 = (IZ25, OZ25, SZ25, NZ25, RZ25)$. Make whatever modifications to the notation you think are necessary. The actual circuit will be connected between $+V_{cc}$ and ground and will contain the necessary biasing components. A model for the mid-frequency behavior of such a transistor circuit is given in the figure (note that the inductor in the base lead is unusual). There are many choices for the state variables. Often it

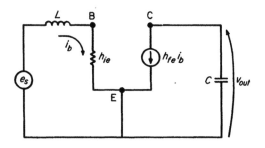

† This problem may require tools not presented in the text.

is wise to associate the state variables with the energy storage elements. Let us choose the current through the inductor, i_b, and the voltage across the capacitor, v_{out}, as our state variables x_1 and x_2, respectively. The state equations become

$$\dot{x} = \begin{bmatrix} \dfrac{-h_{ie}}{L} & 0 \\ \dfrac{h_{fe}}{C} & 0 \end{bmatrix} x + \begin{bmatrix} \dfrac{1}{L} \\ 0 \end{bmatrix} e_s$$

and

$$c^T = (0,\ 1).$$

This problem is analyzed extensively in Szidarovszky and Bahill (1992).

chapter four

Specifying system design requirements

Any system design effort begins with the creation of requirements, which must be stated in a clear, unambiguous way. This chapter discusses what these requirements are and how they should be specified so that a system may be designed to satisfy them.

4.1 The role of systems engineering

Systems engineering differs from most scientific disciplines in that its principal function is problem stating, not problem solving. The problem must be stated in a simple and straightforward manner, without disastrous oversimplification or ambiguity, without confusing ends and means or the abstract with the concrete, without eliminating the ideal solution in favor of the expedient, and without reference to any particular solutions or methods. Needless to say, this is not an easy task.

The statement of the problem should minimize the possibility of something crucial being missed by including a consistent, precise, and comprehensive checklist. The requirements must be defined, including their interdependencies (or lack thereof). The statement of the problem must provide a context within which all solutions can be evaluated and compared, including the "do nothing" solution and bizarre alternatives. Most importantly, the problem must not be stated in terms of a solution, or even a class of solutions.

Why is stating the problem such a difficult task? Perhaps because of the way we are educated. People are taught to think in terms of solutions. Ask anyone about a current national problem—energy, equal opportunity, drugs, etc.—and that person will "know" or at least have an opinion about the solution to the problem, with better educated people even more confident that their solution is the correct one. Rarely will a person suggest that the problem is not well understood and that more resources should be spent in stating the problem before a solution is considered.

In advanced technical education courses, students are taught techniques for solving relatively simple problems and are then assigned exercises for

Figure 4.1 Mr. Wrong Wrench.

practice in doing this. Students emerge from this educational experience as professionals with a tool box mentality—Mr. Wrong Wrenches going about the world looking for problems that can be solved with the tools in their tool boxes and, unfortunately, attempting to solve vastly more complex problems than their simple tools were ever designed for. Many apply their tools without any idea that their tools were developed for problems much simpler than the ones at hand. Furthermore, students are often taught how to use these tools, but not how to decide when a tool is appropriate or how to select between alternative tools; nor do they learn to develop and use intuition, heuristics, rules of thumb, or experience.

Stating the problem is made more difficult by customers who do not know what they want. This is often the result of the customer's preconceived notions or lack of knowledge about what is available. Flexible designs and fast prototyping will help ameliorate this problem, but the fastest way to pin down the customer's desires is to formally describe the system requirements.

The process of defining requirements in detail is illustrated in Exhibit 4.1. The top level system function is to ensure the life cycle satisfaction of requirements. Seven steps are necessary to achieve this function, the first being "develop system requirements." This step can be broken down into five further steps, including "explore problem situation," which is also broken down into further steps. This exhibit shows that a system function consists of subfunctions that, in turn, can consist of further subfunctions. Requirements are developed and expanded by describing these function levels in increasing detail. First a top level function is obtained, and from this function, more detail and information is used to create the additional requirements.

The rest of this chapter describes methods that help the systems engineer state the problem, starting with the top level system function and continuing on in increasing detail.

EXHIBIT 4.1

Systems Engineering Tasks Throughout the System Life Cycle with Details for Phase 1

1. Develop system requirements
    1.1. Explore problem situation
        1.1.1. Record project history
        1.1.2. Research existing systems
        1.1.3. Identify the customer
        1.1.4. Explore system environment
    1.2. Design systems engineering management plan
    1.3. Understand customer's operational need
    1.4. State system requirements
    1.5. Validate system requirements
2. Plan overall design of the system
3. Coordinate detailed design of system components
4. Coordinate acquisition of real system components
5. Coordinate system integration and test
6. Support and monitor system operations
7. Recommend retirement and replacement of system

4.2   *The system design problem*

A system design problem should be stated in terms of the following requirements:

- Input/Output and Functional Requirement,
- Technology Requirement,
- Input/Output Performance Requirement,
- Utilization of Resources Requirement,
- Trade-Off Requirement, and
- System Test Requirement.

Each of these requirements must be defined in detail before a system can be optimally designed. Failure to do this properly will almost guarantee a non-optimal system design.

4.2.1   *Input/Output and Functional Requirement*

The Input/Output and Functional Requirement (IOR) consists of definitions of the time scale, the set of all admissible inputs over time, the set of all eligible outputs over time, and the required functional relationship between the inputs and the outputs.

The IOR is formally defined as follows:

```
IORP0 = {TRP0, IRP0, ITRP0, ORP0, OTRP0, MRP0}
```

The P0 at the end of each acronym represents the system design problem number 0. This can change with the design problem. For the Pinewood case study (see Chapter 5) we use P0, and for SIERRA (see Chapter 6) we use P1. If we had a large system to decompose into smaller problems we would use the number to indicate the decomposition. For example, the entire space shuttle system could be P0, which could then be broken down to the space shuttle, P1, the launch pad, P2, and the solid rockets, P3. The shuttle could in turn be broken down to the main engine, P1.1, cargo bay, P1.2, air frame, P1.3, etc.

TRP0 is the time scale of the system. It explains the units in which time will be measured and defines the system life. It is important for building the system model.

IRP0 is the set of all possible system inputs.

ITRP0 is the set of allowable system input trajectories, which are the inputs as organized in time. This set is often a restriction of all possible input trajectories.

ORP0 is the set of all possible system outputs.

OTRP0 is the set of allowable output trajectories, which are the outputs as organized by time. This set is often a restriction of all possible output trajectories.

MRP0 is the system matching function. It specifies the relationships required between system input trajectories and output trajectories. In the SIERRA case study (Chapter 6), the matching function requires the power to be turned on if either train is detected leaving the danger zone. A formal statement of this is: Given an input for some time element, if that element is (0,1,0,0), i.e., Switch 2 activated, or (0,0,0,1), i.e., Switch 4 activated, then the next output is (1,1), i.e., power on to both trains. This is expressed as:

```
MRP1 = {(f,g):  f ∈ ITRP1;  g ∈ OTRP1;
           if t ∈ TRP1 and
              f(t)=(0,1,0,0) OR f(t)=(0,0,0,1), then
              g(t+1)=(1,1)}.
```

The interpretation: The matching function for Problem 1 (MRP1) is a function of f and g, where f is an element of the input trajectories, and g is an element of the output trajectories. Both are functions of time t. For a given time element t, if the input trajectory is (0,1,0,0) or (0,0,0,1), then the output for the next time element is (1,1).

4.2.2 Technology Requirement

The Technology Requirement consists principally of limitations specified by the customer on the technologies available to build the system. It can include a list of certain components or processes that may or may not be used to solve the problem. It may also include budgets and schedule constraints for the design. Those items that are available for use must be listed in complete textual detail. For example, if all metals must conform to the American Standards for Metals (ASM) specifications, then a reference to ASM parts is needed. Specific design or manufacturing techniques often need be specified as part of the technology requirements.

4.2.3 Input/Output Performance Requirement

The Input/Output Performance Requirement specifies how well the Input/Output and Functional Requirement will be met. The Performance Requirement is expressed in terms of expected response time, expected quality of the response, the number of times an event must occur, etc. These numbers are called figures of merit and must be measured for each system. All of the figures of merit are combined into an overall Input/Output Performance Figure of Merit. The Performance Index is an overall measure of how well the system concept satisfies the requirements.

Figures of merit can be thought of as measurements of quality. They are specific items that need to be quantified to determine whether the concept under study satisfies the design requirements. The System Test Requirement explains how these measurements are carried out. The importance of each figure of merit is rated relative to the others by the customer and systems engineer. A scale of 1 to 10 is usually used for the comparisons. These importance values are then normalized, and the relative weight of each figure of merit is determined. Table 4.1 gives an example of weightings used on the SIERRA case study presented in Chapter 6.

Input/Output Performance Requirements should be broken into classes with three to seven topics each. Each of these topics may also need to be broken down into useful measurements. For example, a top level performance requirement of an electronic device called Reliability could be broken

Table 4.1 Typical weights for the figures of merit.

Requirements	Importance Values, 1 to 10	Relative Weights, IWiP1
1. Number of collisions	8	0.258
2. Trips by Train A	7	0.225
3. Trips by Train B	7	0.225
4. Spurious stops by A	3	0.096
5. Spurious stops by B	3	0.096
6. Availability	2	0.064
7. Reliability	1	0.032

down into sublevels of Junction Temperature, Parts Stress Analysis, and
Number of Parts, as shown below.

1. Reliability
    1.1. Junction Temperatures
    1.2. Parts Stress Analysis
    1.3. Number of Parts

Importance values and weightings for the three items on the sublevel are
determined with respect to each other, but separately from the other levels.

Scoring functions are used to scale different figures of merit to values
between 0 and 1. Calculation of the standard scoring functions (SSF) are done
based on values for the upper, lower, baseline, and slope parameters entered.
We have generalized the 18 scoring functions of Wymore (in press) into one
general purpose scoring function. The four basic shapes that can be derived
from our scoring function are shown in Figure 4.2. We use a value of infinity
for the upper threshold when there is no upper limit and negative infinity for
the lower threshold when there is no lower limit. In general we define the
function SSF to be:

$$\text{SSF (Lower, Baseline, Slope, FigureMerit)} = \cfrac{1}{1 + \left(\cfrac{\text{Baseline} - \text{Lower}}{\text{FigureMerit} - \text{Lower}} \right)^{\text{Power}}}$$

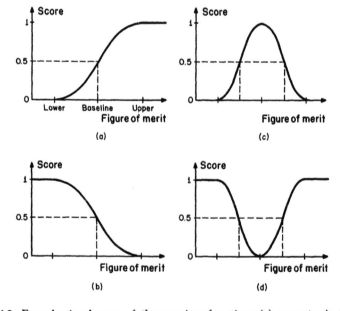

Figure 4.2 Four basic shapes of the scoring function: (*a*) monotonic increasing,
(*b*) monotonic decreasing, (*c*) biphasic hill-shaped, and (*d*) biphasic bowl-shaped.

where

> Lower = lower threshold of the figure of merit,
> Baseline = baseline value of the figure of merit,
> Slope = slope of the curve at the baseline value,
> FigureMerit = value of the figure of merit measured during testing or analysis, and
> Power = 2 * Slope * (Baseline + FigureMerit – 2 * Lower).

The monotonic increasing function of Figure 4.2*a* has the following four ranges:

1. For FigureMerit values less than the lower threshold, Score = 0.
2. For values of FigureMerit between the lower threshold and the baseline value:

$$Score = SSF(Lower, Baseline, Slope, FigureMerit)$$

3. For values of FigureMerit between the baseline value and the upper thresholds:

$$Score = 1 - SSF(2 * Baseline - Upper, Baseline, Slope, FigureMerit)$$

where Upper is the upper threshold of the figure of merit.

4. For FigureMerit values greater than the upper threshold, Score = 1.

For example, in SIERRA the scoring function similar to that shown in Figure 4.2*a* was created to evaluate the number of trips for the model trains. In this case, the lower limit was set to 0, and the upper limit to infinity. The baseline, or expected, value of FigureMerit was 0.5. The baseline parameter indicates the figure of merit that yields a score of 0.5. The slope, measured at the baseline, showed how quickly the score changed at that point. The SSF accepted the value of the observed number of trips and returned a scaled score between 0 and 1. A score of 0.917 was returned for eight trips, and a score of 0.982 was returned for ten trips.

The monotonic, decreasing function of Figure 4.2*b* has a negative slope and is given by the following equation:

$$Score = 1 - SSF(Lower, Baseline, -Slope, FigureMerit)$$

The biphasic functions of Figures 4.2*c* (hill-shaped) and 4.2*d* (bowl-shaped) are pieced together from the monotonic functions of Figures 4.2*a* and 4.2*b*. The upper limit for the first curve must match the lower limit for the second. The program in the appendix (written using the C language) implements these scoring functions.

Figures of merit are measured, and their values are entered into the scoring function to obtain a scaled score. The weights assigned to each figure of merit are multiplied by the scaled score and summed to give the overall Input/Output Performance Index.

More complex methods for comparing systems exist, such as multi-objective analysis. These are beyond the scope of this text, but many good books are available on the subject. See, for example, Szidarovsky, Gershun, and Duckstein (1986).

4.2.4 Utilization of Resources Requirement

The Utilization of Resources Requirement specifies how well the Technology Requirement must be met. It is expressed in terms of expected capital required, schedule constraints, expected operations and maintenance costs, expected environmental and social costs, penalties or rewards for specific component use, etc. All the figures of merit are combined into a Utilization of Resources Index. The weightings and scoring function methods described under the Input/Output Performance Requirement are the same.

4.2.5 Trade-Off Requirement

The Trade-Off Requirement specifies how the Input/Output Performance Requirement is to be weighted with respect to the Utilization of Resources Requirement. The Trade-Off Requirement (TF0) is a weighted sum of the overall Input/Output Performance Figure of Merit (IF0) and the overall Utilization of Resources Figure of Merit (UF0).

The trade-off is made between performance requirements and utilization of resource requirements. In the SIERRA case study (Chapter 6) both were weighted equally, a value of 0.5 being assigned to both TW1P1 and TW2P1. The following equation was used for the trade-off analysis:

$$TF0P1 = 0.5 * 0.801 + 0.5 * 0.989 = 0.895$$

or more generally,

$$IF0P1 = IS1P1(IF1P1) * IW1P1 + \ldots + ISnP1(IFnP1) * IWnP1$$

$$UF0P1 = US1P1(UF1P1) * UW1P1 + \ldots + USmP1(UFmP1) * UWmP1$$

$$TF0P1 = IF0P1 * TW1P1 + UF0P1 * TW2P1$$

where

n = number of Input/Output Performance Figures of Merit,
m = number of Utilization of Resources Figures of Merit,
IF0P1 = overall Input/Output Performance Requirement score,
IFnP1 = n^{th} Input/Output Performance Requirement score,
IWnP1 = weight of the nth Input/Output Performance Figure of Merit,
UF0P1 = overall Utilization of Resources Requirement score,
UFmP1 = m^{th} Utilization of Resources Requirement score,

UW*m*P1 = weight of the *m*th Utilization of Resources Figure of Merit,

UW1P1 = weight of the Overall Input/Output Performance Figure of Merit in the Trade-Off Requirement,

TW2P1 = weight of the Overall Utilization of Resources Figure of Merit in the Trade-Off Requirement, and

TF0P1 = overall score for system design concept.

The Trade-Off Figure of Merit is a mathematical sum, any use of which must include a discussion on the sensitivity of the design to changes.

To perform the sensitivity analysis needed during Concept Exploration in Document 5, one must change the relative weights, figures of merit, or parameters of the scoring function and then recompute the new trade-off scores. These new scores are compared with the previous ones to determine how sensitive the design is to small changes in data. For example, Table 4.2 shows approximate, or blue-sky, guesses for the number of trips completed by each train.

The estimates for figures of merit are put into the scoring function with the values of upper limits, slopes, etc. described above. With these estimates, the overall I/O Performance Index is 0.963. We then play a "what-if" game and ask what happens if we change the approximated values for the number of trips for Train A and Train B. The results are in Table 4.3.

The increased number of trips results in a higher scaled score that, when multiplied by the appropriate weight and summed, yields a score of 0.987. This sum was the basis for computing the trade-off and determining the best system. In other words, a design that allowed each train to make more trips had an overall higher score.

4.2.6 System Test Requirement

The System Test Requirement specifies how, and to what extent, the system that is finally built will be observed and tested in order to determine:

Table 4.2 Approximate values for I/O Figures of Merit.

| | Figure of Merit | | |
Requirements	Value	Score	IWiP1
1. Number of collisions	0	1	0.258
2. Trips by Train A	8	0.917	0.225
3. Trips by Train B	8	0.917	0.225
4. Spurious stops by A	0	1	0.096
5. Spurious stops by B	0	1	0.096
6. Availability	1	1	0.064
7. Reliability	1	1	0.032

Overall Performance Index = 0.963

Table 4.3   Revised approximate values for I/O Figures of Merit.

Requirements	Figure of Merit Value	Score	IWiP1
1. Number of collisions	0	1	0.258
2. Trips by Train A	9	0.961	0.225
3. Trips by Train B	10	0.982	0.225
4. Spurious stops by A	0	1	0.096
5. Spurious stops by B	0	1	0.096
6. Availability	1	1	0.064
7. Reliability	1	1	0.032

Overall Performance Index = 0.987

1. the compliance of the system with all the requirements,
2. the conformance of the system to the design from which it was built, and
3. the acceptibility of the final system to the customer.

The System Test Requirement includes specifications for estimating or measuring values for all the figures of merit defined as part of the Input/Output Performance, Utilization of Resources, and Trade-Off Requirements. These estimates are made on the basis of data collected by blue-sky approximations, by simulations, or by testing the prototypes, system models, or final systems.

The tests that are conducted are based on the inputs defined in test trajectories, which are input trajectories used to determine the figure of merit. Different input trajectories are usually needed to determine the robustness of the system design. For example, in SIERRA the first test trajectory ensures that the trains do not collide when Train A enters the danger zone first, and the second test trajectory ensures that the trains do not collide when Train B enters the danger zone first. These two tests, along with the other trajectories, allow us to calculate the Input/Output Performance Figures of Merit for the Number of Collisions and Number of Trips for Train A and for Train B.

Often these test trajectories are described in terms of scenarios, which are sequences of events. For example, a scenario for a new airplane could begin with the takeoff in conditions of 40-mph crosswinds and a visibility of 200 feet. The scenario continues with a climb to 32,000 feet within 10 minutes, then a level flight for 30 minutes, ending with a night landing on a short runway. The scenario represents a combination of events that tests the entire system, not a single test trajectory for testing a single requirement.

"I've got it, too, Omar...a strange feeling like
we've just been going in circles."

Figure 4.3 Tests must be designed to ensure that systems perform correctly. (The Far Side cartoon by Gary Larson is reprinted by permission of Chronicle Features, San Francisco, CA.

4.3   *General system design*

Customers frequently suggest design requirements that do not fall into a strict category, though they must still be used. In practice, these requirements are usually not broken down strictly either. It is not unusual to see system specifications that mix Technology Requirements with Input/Output Performance Requirements. For example, a specification may state that air temperature is an input to the system and that it must be measured to an accuracy of ±0.1 °C using a digital thermometer. The input to the system is an input/output requirement and the technology limitation is the digital thermometer. The tolerance of ±0.1 °C is a limitation on the system design that may have no meaning. A system could possibly be designed without this restriction, but since it exists in the specification, a non-optimal design may result.

Legal constraints are other important restrictions against function and technology. Many products must comply with national and local laws that the customer may not know. In this case, the government becomes a customer of the system, and careful attention must be given to government rules and regulations. This can impact trade-off decisions profoundly.

Example 4.1

An inventor in New Mexico developed glasses for blind people. These glasses used sonar technology to beep when objects got close. All who tested the product were ecstatic, but lawyers strongly recommended that the product not be built because of future liability considerations—if a person wearing the glasses were to get in an accident, the inventor would undoubtedly be liable by government law. The project was dropped without further consideration.

Environmental conditions are also commonly overlooked by systems designers. The operating, storage, and transportation of a product will place many restrictions on the system design. Most of these restrictions are technology constraints; however, functional design may compensate for environmental considerations by restricting the performance of a system within certain conditions or by modifying the functions, based on the environment.

Example 4.2

A thermocouple does not produce a linear signal over all temperature ranges. An inexpensive thermocouple may be linear only from 20 to 40 °C, but a more expensive device may be linear from 0 to 100 °C. Assume that a system requires a device to function from 10 to 50 °C. In this case, the system requirement is the functional temperature range. The technology requirements are the two available thermocouples. The solution is to either use the more expensive device or possibly adjust the system's function to work with the nonlinear extreme ranges of the cheaper thermocouple. The trade-off between these two options would be determined based on all the factors affecting Input/Output Performance Requirements and the Utilization Of Resource Requirements.

Many complex systems are not documented rigorously at the beginning of the system design because of the uncertainty of the system's success. This is not the best way to design complex systems. Many companies will provide only minimal guidance to system designers believing they are enhancing the creativity of the designers. The result is often a system that does not meet the customer's cost or performance goals, but has lots of bells and whistles that the designers liked.

Example 4.3

Trying to capture the "gee-whiz" market, a major car manufacturer computerized all radio, environmental, and lighting controls in a prototype car. The driver had to access at least two computer screens on a CRT built into the driver's console to make any adjustments. How do you think the consumers responded? Why do you think they responded that way?

Consumers hated it for two reasons:

1. The driver had to look at the screen to activate anything. Most drivers know where the controls in their car are and, keeping their eyes on the road, locate them by touch.

2. Only the driver had comfortable access to the radio, but passengers typically like to adjust the radio.

This is a classic case of the "Voice of the Engineer" driving the system design rather than "Voice of the Customer."

The best way to evaluate a system is to set up measurements of the expected performance of the system and the use of available resources. These measurements can be called figures of merits, quality indicators, product characteristics, or value measures. Whatever they are called, it is important that standards be established against which a design may be evaluated in an unbiased manner. The test requirement determines how these measures are taken. Some companies call these acceptance requirements. Standards for conducting measurements are also important in order to obtain consistent results. Such standards would help eliminate the problem of the "Voice of the Engineer."

Benchmarking competitors is a good way to determine the important measurements. A competitor gaining market share, or solidly in the lead, must have some advantage. A benchmark is a measure of performance or utilization of resources that is taken of your product and that of your competitors. Ideally, the same method of measurement should be used and the measurements should be taken in the same time frame.

Example 4.4

In the early 1980s, the Ford Motor company did not believe there was a quality difference between their cars and foreign imports. To disprove the growing reports of quality problems by the public, Ford obtained the maintenance records of rental car companies. Much to their surprise, the rental agencies were spending twice as much time maintaining U.S. made cars compared to the imports. This benchmark created a major change at Ford that now has it the leading American auto maker in terms of quality.

4.4 Summary

Requirements definition is a critical step in the design process. Requirements that are defined completely and unambiguously in terms the designers understand will result in a system that will satisfy the customer.

The next two chapters are case studies, the first of a Pinewood Derby and the second of a train controller. By following these case studies, one will gain a much clearer understanding of the requirements definition and the use of modeling tools.

Problems

1. **Systems engineering.** Explain the major difference between systems engineering and applied mathematics, applied probability, applied statistics, management, operations research, computer science, human factors engineering, software engineering, electrical engineering, mechanical engineering, manufacturing engineering, and "just *good* engineering."

2. **Scoring functions.** In the Pinewood Derby documentation (Chapter 5), the bell-shaped scoring function of Figure 4.2c was used for the figure of merit for the cost of the new components to be purchased. Why do you think this was done? After all, isn't the cheapest alternative always the best?

3. **Matching functions.** Sometimes we cannot make a neat description of the desired input/output behavior, as we have done so far in this book. In these cases, we need a new tool. With it we describe all possible input trajectories, all possible output trajectories, and a matching function that says which output trajectories are acceptable for each input trajectory.

The figure shows an input trajectory and six possible output trajectories for an automobile windshield wiper system. The input trajectory shows that the switch is on for a while, then off for a while, then on again. The first output trajectory shows that the wiper is stationary. The second output trajectory shows that after receipt of the off input, the wiper stops at its first return to the zero position. In the third output trajectory, after receiving the off input, the wiper completes one more full cycle before coming to rest at zero. In the fourth output trajectory, upon receipt of the off input, the wiper simply stops wherever it is. In the fifth output trajectory, after receiving the off input, the wiper makes a quick return to zero. In the sixth output trajectory, the wiper

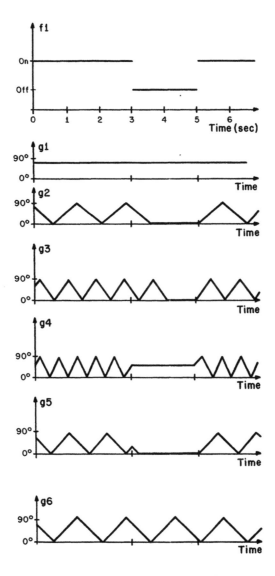

never stops. Write a matching function that pairs the input trajectory to the output trajectories that you think are acceptable. [Based on Wymore (1976).]

Suppose your customer now introduces a new input/output requirement: The wiper frequency must be greater than one cycle per second. What is your new matching function?

4. **Input/Output Requirements–1.** Define an Input/Output Requirement for a system that accepts integers as inputs and produces a real number output that is the average of the last three inputs. Ignore startup transients (i.e., $t = 0, 1, 2$).

5. **Input/Output Requirements–2.** Define an Input/Output Requirement for a system that:

(i)   accepts real numbers or NIL as inputs,

(ii) produces real numbers or NIL as outputs, and

(iii) has a next output of $3x^2 + 2x + 1$ when the input is x; but when the input is NIL, has a next output of NIL.

6. **Input/Output Requirements–3.** Define an Input/Output Requirement for a system that:

(i)   accepts real numbers or NIL as inputs,

(ii) produces real numbers or NIL as outputs, and

(iii) when the input is x, within three time periods the output must be $\sin(x)$ if that is within 10^{-6} of the correct answer; at other times it outputs NIL.

7. **Cooking food.** Assume that your boss just moved into a new house that has an unfinished kitchen, but a refrigerator and pantry full of food. To make the food easier to chew, to kill germs, and to satisfy cultural norms, the food should be cooked. Order a system to cook your boss's food. Make sure you get it right because she is hungry—if you order something that does not work she will probably get mad. At the least, you should specify the Deficiency and the following requirements: Input/Output, Technology, Performance, Utilization of Resources, and System Test. List at least two dozen candidate systems that, at least partially, satisfy your requirements. Your answer should not take more than two pages.

8. **A bridge across the Santa Cruz River.** The following is an edited transcript of an apocryphal meeting between a government administrator (G) and an engineer (E).

G: The bridge across the Santa Cruz River near Mission San Xavier del Bac has been washed out. We want to replace it. As you know, Father Kino, a Jesuit priest the Indians called "the padre on horseback," came to the Indian village named Bac in 1692 and started the first mission. The present church was built in the 1770s. The name of this mission is unusual, as it is a combination of Spanish (Saint Xavier of the) and Indian (place where the water appears) languages. The Santa Cruz River is also very unusual because of its south to north flow. Get me some bids for this bridge.

E: Do you want a wood, steel, or concrete bridge?

G: I don't know. What's the difference?

E: Cost, maximum weight, maximum speed, amount of traffic, system life, and deterioration from rain, salt, and termites. How wide is the river?

G: About 30 meters.

E: How much water flows in the river?

G: Well, none from October to June, but in half the days of July and August it gets one meter…sometimes we see two meters. In September, the average is about 15 centimeters.

E: So, when it rains south of Tucson, there will be water in the river. Otherwise, it is dry?

G: Yes

E: Who uses this road?

G: December to March, hundreds of tourists use it daily. Throughout the year, dozens of Indians use it daily to get from their homes near the mission to their Bureau of Indian Affairs Medical Facility.

E: If there were no bridge, what would be their alternative?

G: A 6 kilometer detour.

E: Is it needed by emergency vehicles such as fire trucks or ambulances?

G: No.

Assume that you are going to do the systems engineering for this project. Write the requirements for this system. At the least, you should specify the Deficiency and the following requirements: Input/Output, Technology, Performance, Utilization of Resources, and System Test. List at least ten candidate systems that you think should be considered. Do not restrict yourself to bridges—often the customers do not know what they want. As with all real world problems, if you do not have data, guess. Your answer should not take more than two pages.

9. **Trade study.** A primitive digital computer system has three operations that must be completed in a certain order. For example, if we wish to add two numbers, we can

A: move x_1 to D0
B: add x_2 to D0
C: move D0 to memory

This system should receive a completion pulse from each subcycle and a check pulse, K, when the major cycle is complete. None of these pulses will occur simultaneously. After the third pulse is received, your system should output a 0 if the pulses occurred in the correct order, or a 1 if they were in an incorrect order. This output will remain until the K pulse resets the system to a state where it is ready to begin another cycle. Each completion signal will occur exactly once in each major cycle. The K check pulse will not occur until after three pulses on A, B, and C have occurred.

Let the Input/Output Requirement, IOR8, be defined as follows:

```
IOR8 = (TR8, IR8, ITR8, OR8, OTR8, MR8), where
TR8 = IJS[0-4],

/*The system must be tested over time {0, 1, 2,
3, 4}.*/

IR8 = {A, B, C, K},
ITR8 = {f1, f2, f3}, where
f1 = STEP(CNS(A), 1, CNS(B), 2, CNS(C), 3,
          CNS(K), 4),

/*This notation says that the input remains
constant with value A up to but not including
time 1, then it holds constant at value B up to
but not including time 2, then it holds constant
at value C up to but not including time 3, and
finally it holds constant at value K up to but
not including time 4. It is a series of step
functions. The figure shows this function
graphically.*/

f2 = STEP(CNS(B), 1, CNS(A), 2, CNS(C), 3,
          CNS(K), 4), /*i.e., (B, A, C, K)*/
f3 = STEP(CNS(A), 1, CNS(C), 2, CNS(B), 3,
          CNS(K), 4), /*i.e., (A, C, B, K)*/
OR8 = {0, 1},
```

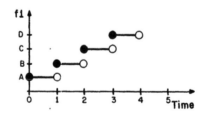

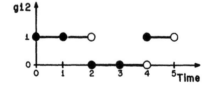

```
OTR8 = {g11, g12, g13, g14, g21, g22, g31, g32,
        g33, g34}, where
g11 = STEP(CNS(0), 5),
      /*i.e., 5 zeroes in a row (0, 0, 0, 0, 0)*/
g12 = STEP(CNS(1), 2, CNS(0), 4, CNS(1), 5),
      /*i.e., (1, 1, 0, 0, 1) in that order*/
g13 = STEP(CNS(0), 1, CNS(1), 2, CNS(0), 5),
      /*i.e. (0, 1, 0, 0, 0)*/
g14 = STEP(CNS(1), 1, CNS(0), 4, CNS(1), 5),
      /*i.e., (1, 0, 0, 0, 1)*/
g21 = STEP(CNS(0), 1, CNS(1), 4, CNS(0), 5),
      /*i.e., (0, 1, 1, 1, 0)*/
g22 = STEP(CNS(1), 5),
      /*i.e., (1, 1, 1, 1, 1)*/
g31 = STEP(CNS(0), 2, CNS(1), 4, CNS(0), 5),
      /*i.e., (0, 0, 1, 1, 0)*/
g32 = STEP(CNS(1), 5),
      /*i.e., (1, 1, 1, 1, 1)*/
g33 = STEP(CNS(0), 1, CNS(1), 4, CNS(0), 5),
      /*i.e., (0, 1, 1, 1, 0)*/
g34 = STEP(CNS(1), 1, CNS(0), 2, CNS(1), 5),
      /*i.e., (1, 0, 1, 1, 1)*/},

/*The next equation says that MR8 is a function
of f and G, where f is an element of the set
ITR8; and G is a subset of OTR8; and G is
further restricted by the following if-then
statements.*/

MR8 = {(f, G): f ∈ ITR8; G is a subset of OTR8;
        if (f = f1) then G = {g11, g12, g13, g14};
        else if (f = f2) then G = {g21, g22};
        else G = {g31, g32, g33, g34}}. /*(f=f3)*/
```

Assume there is no technology restriction.

Find four alternative designs that satisfy the requirements, or in other words, draw state diagrams, specify the systems (i.e., state SZ, IZ, OZ, NZ, and RZ), and specify the system coupling recipes (if necessary) for four systems that satisfy the above input/output requirements.

For the four systems you just designed, assume that our Input/Output Functional Performance figure of merit for the systems that satisfy IOR8 determines that outputs of 1 are expensive and are to be minimized. Which of your designs is best? Assume that the most common input sequence is (A, B, C, K) and that it occurs 90% of the time. Each of the other two input sequences occur 5% of the time. Provide quantitative data. Which scoring function from Figure 4.2 would you use for this data?

10. **Stating the problem.**   A politician asks an engineer, "Can you design a system to solve our water problem?" The engineer replies, "What is the water problem?" The politician says, "In the year 2000 we will have more people than there is water available to keep them alive." The engineer declares, "Shoot some of the people." Flabbergasted, the politician exclaims, "That's not a solution to the problem!" The engineer replies, "It's a solution to the problem that you gave me."

Discuss other examples of social problems that are not well stated and, therefore, generate lots of voluble discussion and misunderstandings.

chapter five

Pinewood

Over 80 million people have participated in Cub Scout Pinewood Derbies. Pinewood is a case study of the design of a Cub Scout Pinewood Derby for one particular scout pack. The system helps manage the entire race from initial entry through final results. Many alternatives for race format, scoring, and judging are presented.

The following detailed table of contents should be examined closely by the systems engineer. Our computer program requires that the designers provide an entry for every table item. This will ensure that the important points are not overlooked. .

Note that in the seven documents that follow many comments are set in italics, indented, and bounded by a box (such as this one). They are not a part of the system documents, but are comments for the reader. They contain explanations and indications of the strong and weak points of this documentation.

Individual Cub Scout packs usually hold their Pinewood Derbies at the end of January or the beginning of February; district and city-wide derbies follow. You should find out when and where a Pinewood Derby will be held in your neighborhood and attend it.

Contents

5.1 Document 1: Problem Situation

The Problem Situation Document is the executive summary. It explains the problem that needs to be solved. It is written in plain language and is intended for management.

5.1.1 The top level system function

The top level system function is to conduct a cub scout Pinewood Derby that maximizes scout enjoyment and minimizes hard feelings.

5.1.2 History of the problem and the present system

Since the 1950s, over 80 million cub scouts have built cars and raced them in Pinewood Derbies. Pack 212 in Tucson, Arizona, has conducted derbies since 1977. Problems that have developed in past Pinewood Derbies include:

1. scouts and parents wasting large amounts of time,
2. irate parents,
3. questions about the fairness of races,
4. other people touching the scouts' cars,
5. adverse weather conditions,
6. scouts unable to tell which cars were called to race or in which lane the cars were to run,
7. scouts unable to tell which cars won, and
8. scouts unable to figure out which cars were winning the derby.

The cub scouts build cars from a Pinewood Derby Kit to prescribed requirements. Systems engineers will design a derby to alleviate the existing adverse factors. This project is known as Pinewood.

5.1.3 The customer

5.1.3.1 Owners
The system will be owned by Cub Scout Pack 212, Catalina Council, Boy Scouts of America.

5.1.3.2 Bill payers: The client
The budget for the system will be provided by Dr. A. Terry Bahill.

5.1.3.3 Users
The system will be used by the cub scouts of Pack 212, their parents, and the Pinewood Derby Committee.

5.1.3.4   Operators
The system will be operated by the members of the Pinewood Derby Committee (judges, inspectors, track managers, etc.) of Pack 212.

5.1.3.5   Beneficiaries
The cub scouts, their parents, the organizers, and spectators are the beneficiaries of the system.

5.1.3.6   Victims
Those who might feel the system adversely affected them are

1. those cub scouts who lose,
2. cub scouts whose cars are broken,
3. disgruntled parents,
4. those who must clean up the area after the event, and
5. committee members who take verbal abuse from irate parents.

5.1.3.7   Technical representatives to systems engineering
The sole technical representative of this system is the system designer, Dr. Bahill of the University of Arizona.

5.1.4   Technical personnel and facilities

5.1.4.1   Life Cycle Phase 1:   Requirements development
Dr. Bahill is the technical consultant for the basic system throughout Phase 1. All requirements data will be supplied by Dr. Bahill. Supplies and tools will be provided by Dr. Bahill. Computer equipment for document generation will be provided by the system designers.

5.1.4.2   Life Cycle Phase 2:   Concept development
The system designers will perform the concept development and will be available throughout Phase 2. Information resources obtained from previous derbies will be provided by Dr. Bahill. Computer resources for simulations will be provided by the system designers. Bill Chapman will be the systems engineer.

5.1.4.3   Life Cycle Phase 3:   Full-scale engineering development
The full-scale engineering task will be performed by Dr. Bahill and Bill Karnavas. A three lane racetrack will be provided by the cub scout pack. Computers and timing hardware will be provided by Dr. Bahill.

5.1.4.4   Life Cycle Phase 4:   System development
The system development will be performed by Dr. Bahill and Bill Karnavas.

5.1.4.5   Life Cycle Phase 5:   System test and integration
System test and integration will be performed by Dr. Bahill and Bill Karnavas.

5.1.4.6 Life Cycle Phase 6: Operations support and modification
Following successful system test and integration, operations support and modification will be performed by the Pinewood Derby Committee.

5.1.4.7 Life Cycle Phase 7: Retirement and replacement
At the end of the race day, the system will be disassembled and the equipment will be stored. Next year, a replacement system will be designed and built.

5.1.5 System environment

5.1.5.1 Social impact
The primary social impact of the new system is to provide a better overall derby, which will be more organized, more efficient, and more enjoyable. For the children who race the cars, competition is de-emphasized and racing is emphasized. By this we mean that the format or structure of the event should allow scouts to participate in a large number of races, thus keeping their attention focused on the races. The scouts learn that their own actions, rather than luck, control who wins and loses.

5.1.5.2 Economic impact
The new system will improve the utilization of the economic resources. Though the new system might not require more resources than existing systems, it may cost up to $300 more.

5.1.5.3 Environmental impact
The local environment may be affected by debris from the crowd or by graphite deposits left by the scouts (the scouts use graphite during the races). This will have to be cleaned up following the event. The Pinewood Derby Committee or maintenance personnel will restore the environment to an acceptable state. Other potential problems are noise and parking congestion during the event.

5.1.5.4 Interoperability
The system must be compatible with the environment and the established components of the derby, such as the pinewood cars and the racing track. Furthermore, the system must be in compliance with existing Boy Scouts of America Pinewood Derby specifications.

5.1.6 Systems engineering management plan

The system designers will describe the project design within the seven system engineering documents:

1. Problem Situation Document,
2. Operational Need Document,

3. System Requirements Document,
4. System Requirements Validation Document,
5. Concept Exploration Document,
6. System Functional Analysis Document, and
7. Physical Synthesis Document.

These documents will be continually updated as the design progresses using the SEDSO software package (see Chapter 7 for details on SEDSO). Furthermore, the Pinewood Derby Committee will be responsible for the project from the end of the system test and integration phase to the end of the life cycle.

5.2 Document 2: Operational Need

> *The Operational Need Document is a detailed description of the problem in plain language. It is intended for management, the customer, and systems engineers.*

5.2.1 Deficiency

In the past, the emphasis for this derby was placed on winning, rather than racing. Also, hard feelings were created by wasted time and what the parents and the scouts perceived to be incorrect or unfair judging. The new system will change the emphasis to racing, reduce the number of irate parents, and increase the number of happy kids.

5.2.2 Input/Output and Functional Requirement

5.2.2.1 Time scale
The system will use a time scale with a resolution of tenths of milliseconds. The life expectancy of the system will be six hours.

5.2.2.2 Inputs
The system has eight inputs:

1. name of the owner of the current Pinewood Derby car entering the system,
2. division the owner is in,
3. den the owner is a member of,
4. car's speed ability,

5. car's compliance with derby rules (i.e., pass or no pass),
6. time of day,
7. scheduled judging time for each event, and
8. scheduled racing order for each race.

The divisions are Webelos, Bears, Wolves, Tiger Cubs, and Family. No two owners may use the same name. The dens have separate unique names or numbers. The scouts belong to both a den and a division. The Family cars will not have a den designation. The scheduled racing order will depend on the format of the racing technique, though it will be determined in advance and provided to the system.

> *Family cars are built by fathers, mothers, or siblings obeying the same rules as the scouts. The original purpose of the Family car division was to cajole the fathers into leaving the scouts' cars alone by building cars of their own. This worked quite well, the kids' cars being built by the kids. Subsequently, the Family car division developed an added facet of presenting truly innovative and fancy designs. Some cars were built for speed; some were built for originality, such as a three-wheeled, inchworm-shaped car; and some were built to reflect family occupations—UPS trucks, window glass delivery trucks, etc.*

5.2.2.3 Input trajectories
The system input trajectories will be restricted to the order of divisional racing: Webelos, Bears, Wolves, Tiger Cubs, and Family.

5.2.2.4 Outputs
The outputs of the system are indicators of:

1. the first, second, and third place finisher of each race,
2. the name, division, and den of the first, second, and third place finishers in each division event,
3. the first, second, and third place winners of the Pack Championship and the Family car races and the winner of the Classy Chassis Competition for each division,
4. the first, second, and third place winners of each den and a list of the other den entrants,
5. scouts who are either happy or not,
6. parents who are either irate or not, and
7. qualifying or disqualifying of cars.

5.2.2.5 *Output trajectories*
The output trajectories shall be restricted as follows:

1. The determination of the division winners will precede the Pack Championship race, and the Family car category will conclude the derby events.
2. No race can end in a tie.
3. The final Classy Chassis determinations will occur after all the events are completed.

5.2.2.6 *Matching function*
The required matching between input trajectories and output trajectories are as follows:

1. The Webelos winner will be a car from the Webelos car division.
2. The Bears winner will be a car from the Bears car division.
3. The Wolves winner will be a car from the Wolves car division.
4. The Tiger Cubs winner will be a car from the Tiger Cubs division.
5. One Classy Chassis winner will be selected from the Family cars and one will also be selected from the Webelo, Bear, and Wolf division nominees.
6. The Pack Champions will be from the Webelos, Bears, or Wolves division.
7. The Family cars winner will be from the Family car division.

5.2.3 *Technology Requirement*

5.2.3.1 *Available money*
The time spent by the volunteers, Dr. Bahill and Bill Karnavas, is considered free. Dr. Bahill says that $200 is not an unreasonable amount of out-of-pocket money to spend. The pack will pay gym rental fees, if needed. This will usually cost $25 to $100.

5.2.3.2 *Available time*
The project must be completed before the scheduled date of the derby, which is the first Sunday in February for Pack 212.

5.2.3.3 *Available components*
Available components are

1. an IBM AT computer,
2. timing equipment and software,
3. a stopwatch,
4. committee personnel and other volunteers,
5. a three-lane racetrack,
6. awards and prizes,
7. weighing scales and rulers,
8. tables and chairs, as necessary, and

9. other materials that can be obtained "off-the-shelf," as needed and permitted by the budget.

5.2.3.4 Available techniques

Of the many different racing techniques that can be considered, we will use the following:

1. single-elimination,
2. double-elimination, and
3. round-robin formats with the following scoring techniques:
   3.1. mean times,
   3.2. fastest times, and
   3.3. point assignments.

Many timing techniques are available for determining the order in which the cars cross the finish line. A list of potential techniques includes

1. optical sensors,
2. bar code readers,
3. mechanical switches, and
4. human observation.

5.2.3.5 Required interfaces

The proposed system is required to interface with Pack 212's existing three-lane racetrack and Pinewood Derby cars.

5.2.3.6 Standards, specifications, and other restrictions

The design, implementation, and operation of the system must follow the Boy Scouts of America Pinewood Derby rules and regulations, as described in Exhibit 5.1.

EXHIBIT 5.1

Typical Pinewood Derby Rules

Cars must be built using an Official Cub Scout Pinewood Derby Kit; however, weights, paint, decals, decoration, and graphite may be added. Other wheels or axles are *not* permitted, as we do not want the scouts to buy expensive components. The cars should be built by the scouts using commonly available tools. Thus, wheels may be sanded smooth, as described in the Pinewood Derby Kit, but they may not be turned on a lathe to produce knife edges. Likewise, axles may be smoothed, but they cannot be plated. All parts of the car must be firmly attached. The car must have proper clearance underneath; weights may not be hung under the car. Nothing can project beyond the front of the car. All cars must be built in the year of the derby. The cars should be built by the scouts. Fathers, mothers, brothers, and others may build their own cars and race them in the Family car division races. On race day, each scout should bring his car, graphite, and a tool for reducing the weight of the car if it exceeds five ounces. In addition, cars must also comply with the following council rules:

1. After inspection on race day nothing can be done to the cars. Graphite may *not* be added to the wheels after inspection. In particular, neither scouts nor parents can add graphite to the wheels between races.
2. Width shall not exceed 2.75 inches.
3. Length shall not exceed 7 inches.
4. Weight shall not exceed 5 ounces.
5. Axles, wheels, and body shall be from materials provided in the kit.
6. Wheel bearings, washers, and bushings are prohibited.
7. Wheels and axles may be lubricated with graphite, but oil may not be used.
8. Springs are not allowed.
9. The car must be free-wheeling, and there can be no starting devices.
10. No loose materials are allowed in or on the car.
11. The wheelbase must not be altered from that in the kit.

In some years, the district allows scouts to use expensive, precision-machined wheels bought from mail-order hobby houses. At that time, our Cubmaster will buy such knife-edged wheels for our scouts to use in the district competition.

Helpful hints from Dr. Bahill: In decreasing order of importance, the things that make a Pinewood Derby car go fast are

1. graphite—make sure there is lots of graphite between the wheels and axles;
2. weight—make the car as close to five ounces as possible;
3. smoothness of wheels and axles—sand your wheels and polish your axles;
4. weight distribution—the center of mass should be toward the back of the car, e.g. an inch or so in front of the rear axle;
5. mounting of wheels—put your axles in straight, however it is not necessary that all four wheels touch the ground; and
6. aerodynamics—at these low speeds wind resistance has no effect.

Concurrent engineering requires that all decisions be made with the full participation of all relevant personnel. We document this by indicating the primary originating participant following the figures of merit. Such participants may include sales and marketing, finance, manufacturing, engineering, quality, human factors, and purchasing.

5.2.4 Input/Output Performance Requirement

1. *Average Races per Car:* The average number of races per car. There is no defined upper limit. In 1990, the number of races per scout was six so this is the baseline. This requirement was devised using Sales and Marketing data.

In these requirements, we suggest the divisions of a large company that might be responsible for suggesting each requirement. For the Pinewood Derby this may seem a little contrived, but it does illustrate how concurrent engineering (explained in Chapter 7) works for larger systems.

2. *Number of Ties:* The total number of times that races had to be rerun in the entire derby because of ties. An upper limit of 15 ties has been set with a baseline value of 0.5. This requirement was set by Systems Engineering.

3. *Happiness:* The happiness of the scouts and parents resulting from the derby. This is a combination of the following seven measures:

    3.1. *Percent Happy Scouts:* The percentage of scouts that leave the race with a generally happy feeling. A happy feeling may be the result of a child having a good race, having a good rapport with other scouts and parents, or a combination of these factors. It should be maximized to meet the top level system function. The upper limit is 100%, and 95% is the baseline. This requirement was suggested by Sales and Marketing and the customer.

    3.2. *Number Irate Parents:* The total number of parents that are dissatisfied with the judging of the races or any other aspect of the race. The upper limit is 10 and the baseline value is 1. These criteria were determined by the customer and Human Factors data.

3.3. *Number of Broken Cars:* The number of cars that were broken by the system itself. The upper limit is 3, and the baseline value is 0.5, since we really do not want any cars broken. This is a customer requirement.

3.4. *Others Touching Scout's Car:* The number of other people who touch the scout's car during a race. The upper limit is 7 and the baseline value is 2. This requirement was specified by the customer.

3.5. *Number of Repeat Races:* The number of cars that race another particular car more than once. A larger number of repeat races will increase the perception of fairness and lower the discontent of the scouts. This requirement was made by Human Factors and Systems Engineering.

3.6. *Number of Lane Repeats:* The number of cars that do not race the same number of times in each lane. A smaller number of lane repeats will increase the perception of fairness and lower scout discontent. This requirement was determined by Human Factors and Systems Engineering.

3.7. *Difference Between Fast and Slow:* The difference between the number of races for the fastest car and the number of races for the slowest car. This requirement was determined by Systems Engineering.

Notice how we have grouped related subitems together into one fig-ure of merit, Happiness. It is important to group related items so that individual items do not gain too much importance. We try to keep the number of items at any level between 3 and 7, so comparisons can be made easily.

---

4. *Availability:* The system will be available if it interfaces with the current track system and is manufactured on time and to specification. This requirement was determined by Systems Engineering.

5. *Reliability:* The system will be reliable if it behaves at least as well as the existing system and if it can operate in case of electrical power failure. This requirement was determined by Reliability Engineering.

Some systems engineers do a risk analysis after the most favorable alternatives are selected. We chose to merely incorporate the risk parameters into the requirements. For example, the risk of a total power failure on the day of the race was incorporated into the Reliability Input/Output Performance Requirement.

5.2.5 Utilization of Resources Requirement

1. *Acquisition Time:* The number of days the project was completed before the first Sunday in February. The sooner the system is completed before this time the better. This requirement was determined by the customer and Purchasing.

2. *Acquisition Cost:* The total cost of creating the system. The absolute maximum is $300, and the baseline value is $150. This requirement was determined by the customer and Purchasing.

3. *Total Event Time:* Total time it takes to judge all cars and to run all races. In 1990, the derby took 3.5 hours to complete; so this is our baseline. This requirement was suggested by the customer.

It is not always clear when a figure of merit should be grouped with Input/Output Performance or Utilization of Resources. For the Pinewood Derby, it seems that Total Event Time could go in either category.

4. *Number of Electrical Circuits:* The number of 120 VAC electrical circuits needed to run the event. The baseline value is 1, with an optimum score of 0 circuits. This requirement was determined by Manufacturing.

5. *Number of Adults:* The total number of adults needed to run the derby. This requirement was created by Manufacturing.

5.2.6 Trade-Off Requirement

Pinewood's trade-off analysis gives greater weight to the performance requirements (90%) than to the resource requirements (10%) because the parents want their kids to be happy and they are willing to pay for it. This requirement was created by management.

5.2.7 System Test Requirement

The performance of the system designed by the system engineers will be determined using two tests. These requirements were created by the System Test Organization.

1. Test 1 will determine system performance using 23 cub scouts from each division.
2. Test 2 will determine if the race judging components are fair. Two cars with similar speeds will be used for this. Dr. Bahill and Bill Karnavas will be the judges.

The system will be acceptable if

1. all requirements from this document are satisfied,
2. the system allows for adverse weather conditions,
3. at most 1500 square feet of space are used, and
4. restroom facilities are available for participants.

The system will be in compliance if the upper and lower bounds set for each figure of merit are met. The system will have failed if

1. there is a loss of electrical power and power is needed,
2. adverse weather prevents the derby from proceeding,
3. mistakes in judging occur, or
4. one lane is faster than another.

These will be determined by the Grand Marshall during the actual event.

5.2.8 Rationale for operational need

The data and specifications were provided by Dr. Bahill and Bill Karnavas.

Below are listed some things we actually do for each derby we run but that were omitted from this documentation either because we forgot to include it or because we thought it would needlessly complicate the documentation.

(1) Find out how many scouts are in each division. Obtain historical data for time per race for each division, as shown in Exhibit 5.2. Produce a timetable to minimize wasted time. With electronic timing, we found that we could schedule a race every 45 seconds. Races can be run even faster for older kids and adults. Also, later races can be run faster because the track needs no further adjustment and because the parents have learned their jobs. Small races, with 12

cars or less, do not require impounding of cars between races and thus can be run faster.

(2) Publish car construction rules for the pack two months before the event.

(3) Meet with the Pinewood Derby Committee and explain each person's job.

(4) Provide a listing of who won the various prizes within one week after the derby.

EXHIBIT 5.2

Statistical Summary of the 1991 Pack 212 Pinewood Derby

Pack 212 1991 Pinewood Derby

Division	Number of Cars	Percentage of Scouts Participating	Number of Races	Duration of Event (minutes)	Time Used per Race (minutes)
Webelos	23	62	48	35	0.73
Bears	16	84	36	23	0.64
Wolves	10	77	24	15	0.63
Tiger Cubs	7	87	18	10	0.56
Pack Championship	9		18	10	0.56
Family Cars	10		24	15	0.63
Totals	66		168		

We used an electronic timer and ran a round-robin derby with each car racing six times, twice in each lane. From this summary, we can see that with electronic timing one race every 45 seconds is a reasonable schedule. The first division will be the slowest because of the time taken to cross check the computer and straighten and wax the track. With small numbers of cars per division—that is, 12 or fewer—impounding the cars between races is not desirable, since more races can be run in the same period of time by not impounding them. These statistics are very similar to those of the previous year.

5.3 Document 3: System Requirements

> The Systems Requirements Document is a succinct mathematical description or model of the Input/Output Requirements, Functional Requirements, Technology Requirements, Test Requirements, and the trade-offs between them as described in Document 2. Its audience is systems engineers.

5.3.1 The system requirement

The System Design Problem entails stating the following requirements.

- Input/Output and Functional Requirement,
- Technology Requirement,
- Input/Output Performance Requirement,
- Utilization of Resources Requirement,
- Trade-Off Requirement,
- System Test Requirement.

Each of these requirements will be mathematically stated in the following sections.

5.3.2 Input/Output and Functional Requirement

5.3.2.1 Time scale

TRPO is the time scale of Pinewood expressed in tenths of a millisecond. The life expectancy of the system is six hours. This becomes 6 hours × 60 minutes/hour × 60 seconds/minute × 10,000 = 216,000,000.

```
TRPO = IJS[0-216000000]
```

> This time scale does not presuppose that electronic timing will be used. It was chosen to be fast enough to work with all alternatives. Slower models would certainly be valid.

.5.3.2.2 Inputs

IRPO represents the set of system inputs for Pinewood. There are four input ports:

```
IRPO = IR1PO × IR2PO × IR3PO × IR4PO
```

where `IR1PO` is a set of sets of all possible car entries and is broken down as follows:

```
IR1PO = Carin = {Owner, Den, Division, Speed,
                     Characteristic}
```

where

```
Owner = {Words(Alphau)}
```

"Alphau" is a function that returns any letter or number. "Words" is a function that puts the alphanumerics into a word.

```
Den = {Words(Alphau)}
Division = {Webelos, Bears, Wolves, Tiger Cubs,
              Family}
Speed = IJS[1-100]
Characteristic = {Pass, Fail}
```

`Speed` is a relative measure used for simulation. We do not know how fast the cars are, but they enter the system with some inherent speed capability. Likewise, `Characteristic` represents the car's ability to ultimately `Pass` or `Fail` the inspection. This part of the modeling is simplistic, since we are not interested in an in-depth model of this portion of the system.

`IR2PO` is the time of day provided to the system.

```
IR2PO = IJS[0, 2160000000].
```

`IR3PO` is the scheduled judging times.

```
IR3PO = {Division, Time}
```

where

```
Division = {Webelos, Bears, Wolves, Tiger Cubs,
              Family}
Time = IJS[0-216000000].
```

`IR4PO` is the scheduled racing order.

```
IR4PO = {(Index, Lane1, Lane2, Lane3)^Num}
```

where

```
Index = IJS[0-Num] /*Index is the race number*/
                   /*on the schedule*/
```

```
Lane1 = Carin        /*The car in lane 1*/
Lane2 = Carin        /*The car in lane 2*/
Lane3 = Carin        /*The car in lane 3*/
Num = 1000           /*The max number of possible*/
                     /*races*/
```

5.3.2.3 Input trajectories

ITRPO is the set of input trajectories for Pinewood, the set of all possible inputs (IRPO) over the time scale (TRPO). Formally,

```
ITRPO = {f: f ∈ FNS(TRPO,IRPO);
            f(t) = ((p11(t),p12(t),p13(t),p14(t),
                    p15(t)),p2(t),p3(t),p4(t)),
            tj ∈ TRPO, j={1,2,3,4,5};
            if f(t1) = ((p11,p12,Webelos,p14,p15),
                        p2,p3,p4) and
               f(t2) = ((p11,p12,Bears,p14,p15),
                        p2,p3,p4) and
               f(t3) = ((p11,p12,Wolves,p14,p15),
                        p2,p3,p4) and
               f(t4) = ((p11,p12,Tiger Cubs,p14,
                        p15),p2,p3,p4) and
               f(t5) = ((p11,p12,Family,p14,p15),
                        p2,p3,p4) then
                  t1 < t2 < t3 < t4 < t5}.
```

where, for example, p12 is the second element of the first port and, similarly for all the others of f(t), where f(t) is the resultant input trajectory at time t.

5.3.2.4 Outputs

ORPO represents the system outputs for Pinewood.

```
ORPO = OR1PO × OR2PO
```

where OR1PO is a set of sets of cars as follows:

```
OR1PO = Cars = {Owner, Den, Division, Timein,
                Place, Event, Qual, Scout,
                Parent}
```

where

```
Owner = {Words(Alphau)}
Den = {Words(Alphau)}
Division = {Webelos, Bears, Wolves, Tiger Cubs,
            Family}
Timein = IJS[0-2160000000],
Place = {First, Second, Third, Null}
```

```
Event = {Race, Pack Championship, Classy Chassis}
Qual = {Qualified, DisQualified}
Scout = {Happy, Nothappy}
Parent = {Irate, Notirate}.
```

These outputs indicate conditions of the cars, the scouts, and the parents. Qual is the output that indicates whether the car is, or is not, qualified to race.
OR2PO is a set of sets of cars as follows:

```
OR2PO = Cars = {Owner, Den, Division, Timein,
                Place, Event, Qual, Scout,
                Parent}
```

where Cars is defined as above.

5.3.2.5 Output trajectories
OTRPO is the set of all output trajectories for Pinewood. OTRPO is the set of all possible outputs (ORPO) over the time scale (TRPO). Formally,

```
OTRPO = {f: f ∈ FNS(TRPO,ORPO), and
            for t ∈ TRPO and
            for OR1PO = (q1,q2,q3,q4,q5,q6,q7,q8,
                         q9),
            if q3 = Webelos then t1 = t;
            else if q3 = Bears then t2 = t;
            else if q3 = Wolves then t3 = t;
            else if q3 = Tiger Cubs then t4 = t;
            else if q3 = Family then t5 = t; and
                t1 < t2 < t3 < t4 < t5}
```

where q3 represents the third element of the output set OR1PO, which is the racing division, and t1 is the time when the Webelos race begins.

5.3.2.6 Matching function
MRPO is the matching function for Pinewood.

```
MRPO = {(f,g): f ∈ ITRPO; g ∈ OTRPO, and
            for f=(t1, (p11,p12,p13,p14,p15),
                   p2,p3,p4) ∈ ITRPO, and
            for g=(t2, (q1,q2,q3,q4,q5,q6,q7,
                        q8,q9) ∈ OTRPO then
            if q3 = Webelos then t2a = t2;
            else if q3 = Bears then t2b = t2;
            else if q3 = Wolves then t2c = t2;
            else if q3 = Tiger Cubs then t2d = t2;
            else if q3 = Family then t2e = t2; and
            if p13 = Webelos then t1a = t1;
            else if p13 = Bears then t1b = t1;
            else if p13 = Wolves then t1c = t1;
```

```
else if p13 = Tiger Cubs then t1d = t1;
else if p13 = Family then t1e = t1;
   then t1a < t2a and t1b < t2b and
         t1c < t2c and t1d < t2d and
         t1e < t2e}
```

where, for example, q3 represents the third element of the output set OR1PO, which is the racing division; p13 is the third element of the first element of the input trajectory, which is the car's division; and t1a is the time when the Webelos race begins and t2a is the time the race ends.

5.3.3 *Technology Requirement*

> *Section 5.3.3 is very similar to Section 5.2.3. For material for which mathematical models are not appropriate, the sections of Documents 2 and 3 will be similar, but we do not eliminate one or the other because each document must be self-contained.*

5.3.3.1 *Available money*
Dr. Bahill says that $200 in out-of-pocket expenses is not an unreasonable amount to spend. Gym rentals will cost approximately $25 to $100. If the Pack cannot afford this cost by the time of the event, then the race must be held elsewhere, possibly outside in someone's yard.

5.3.3.2 *Available time*
Though the time spent by Dr. Bahill and Bill Karnavas is a resource that should not be squandered, their time before, during, and after the derby is considered free.

5.3.3.3 *Available components*
The following components are available:

1. an IBM AT computer,
2. timing equipment and software,
3. a stopwatch,
4. committee personnel and other volunteers,
5. the three-lane racetrack,
6. awards and prizes,
7. weighing scales and rulers,
8. tables and chairs, as necessary, and
9. other materials that can be obtained "off-the-shelf," as needed and permitted by budget.

5.3.3.4 Available techniques
Preferred racing techniques include:

1. single-elimination,
2. double-elimination, and
3. round-robin formats with the following scoring techniques:
    3.1. mean times,
    3.2. fastest times, and
    3.3. point assignments.

Candidate timing techniques include:

1. optical sensors,
2. bar code readers,
3. mechanical switches, and
4. human observation.

5.3.3.5 Required interfaces
The proposed system is required to interface with Pack 212's existing three-lane racetrack and derby car sizes.

5.3.3.6 Form, fit, and other restrictions
These considerations include the size of the existing racetrack and the space needed to house all the participants in the event along with all inspection and timing stations. Estimated minimum floor space is 1500 square feet. The event should be held indoors to prevent adverse effects from the weather; otherwise, arrangements for holding the event in good weather should be made.

5.3.3.7 Standards and specifications
The Pinewood Derby system must comply with all rules and regulations of the Boy Scouts of America pertaining to Pinewood Derbies. Also, safety practices and procedures should be followed, and any building rules and codes must be obeyed.

5.3.4 Input/Output Performance Requirement

5.3.4.1 Definition of Performance Figures of Merit
The overall performance figure of merit is denoted IF0P0 and is computed as follows:

$$IF0P0 = ISF1P0 * IW1P0 + ISF2P0 * IW2P0 + \ldots + ISFnP0 * IWnP0$$

where n is the total number of I/O Performance Figures of Merit and

$$ISFiP0 = ISiP0(IFiP0(FSD)) \quad \text{for } i = 1 \text{ to } n$$

as explained in the following section.

5.3.4.2 Lower, upper, baseline, and scoring parameters
In this section, the following naming convention is used: The initial letter "I" indicates that the name is for an Input/Output Performance Requirement. The terminal P0 indicates that the name involves Problem 0 of the Pinewood Derby.

$\text{IF}i\text{P0}$ = the ith figure of merit measured per the test plan,
$\text{IB}i\text{P0}$ = the baseline value for the ith figure of merit,
$\text{IFX}i\text{P0}$ = measured value for the ith figure of merit,
$\text{ILTH}i\text{P0}$ = lower threshold for the ith figure of merit,
$\text{IR}i\text{P0}$ = ranking of importance from 1 to 10,
$\text{ISF}i\text{P0}$ = score for the ith figure of merit,
$\text{IS}i\text{P0}$ = scoring function for the ith figure of merit,
$\text{ISL}i\text{P0}$ = slope for the ith figure of merit,
$\text{IUTH}i\text{P0}$ = upper threshold for the ith figure of merit,
$\text{IW}i\text{P0}$ = weight for the ith figure of merit, and
SSF = standard scoring function.

Next we give the parameters necessary to evaluate the figures of merit using the scoring functions of Figure 4.2.

1. *Average Races per Car*

    Score   IS1P0 = SSF (ILTH1P0,IB1P0,IUTH1P0,ISL1P0)

    Lower Threshold   ILTH1P0 = 1

    Baseline   IB1P0 = 4

    Upper Threshold   IUTH1P0 = ∞

    Slope   ISL1P0 = 0.333

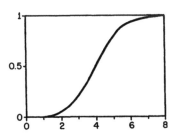

Figure 5.1 Scoring function.

2. *Number of Ties*

    Score   IS2P0 = SSF (ILTH2P0,IB2P0,IUTH2P0,ISL2P0)

    Lower Threshold   ILTH2P0 = 0

    Baseline   IB2P0 = 0.5

    Upper Threshold   IUTH2P0 = 5

    Slope   ISL2P0 = –2

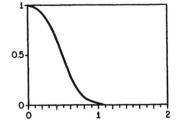

Figure 5.2 Scoring function.

3. *Happiness*

    Score   IS3P0 = SSF (ILTH3P0,IB3P0,IUTH3P0,ISL3P0)

    Lower Threshold   ILTH3P0 = 0

    Baseline   IB3P0 = 0.5

    Upper Threshold   IUTH3P0 = 1

    Slope   ISL3P0 = 2

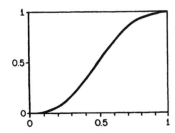

Figure 5.3 Scoring function.

3.1. *Percent Happy Scouts*

Score   IS3.1P0 = SSF (ILTH3.1P0,IB3.1P0,IUTH3.1P0,ISL3.1P0)

Lower Threshold   ILTH3.1P0 = 0

Baseline   IB3.1P0 = 90

Upper Threshold   IUTH3.1P0 = 100

Slope   ISL3.1P0 = 0.1

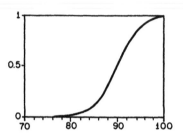

Figure 5.4 Scoring function.

3.2. *Number Irate Parents*

Score   IS3.2P0 = SSF (ILTH3.2P0,IB3.2P0,IUTH3.2P0,ISL3.2P0)

Lower Threshold   ILTH3.2P0 = 0

Baseline   IB3.2P0 = 1

Upper Threshold   IUTH3.2P0 = 10

Slope   ISL3.2P0 = −1

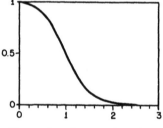

Figure 5.5 Scoring function.

3.3. *Number of Broken Cars*

Score   IS3.3P0 = SSF (ILTH3.3P0,IB3.3P0,IUTH3.3P0,ISL3.3P0)

Lower Threshold   ILTH3.3P0 = 0

Baseline   IB3.3P0 = 1

Upper Threshold   IUTH3.3P0 = 3

Slope   ISL3.3P0 = −1

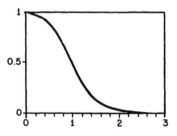

Figure 5.6 Scoring function.

3.4. *Others Touching Scout's Car*

Score $\qquad$ IS3.4P0 = SSF (ILTH3.4P0,IB3.4P0,IUTH3.4P0,ISL3.4P0)

Lower Threshold $\qquad$ ILTH3.4P0 = 0

Baseline $\qquad$ IB3.4P0 = 2

Upper Threshold $\qquad$ IUTH3.4P0 = 7

Slope $\qquad$ ISL3.4P0 = −0.5

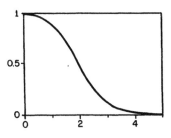

Figure 5.7 Scoring function.

3.5. *Number of Repeat Races*

Score $\qquad$ IS3.5P0 = SSF (ILTH3.5P0,IB3.5P0,IUTH3.5P0,ISL3.5P0)

Lower Threshold $\qquad$ ILTH3.5P0 = 0

Baseline $\qquad$ IB3.5P0 = 2

Upper Threshold $\qquad$ IUTH3.5P0 = ∞

Slope $\qquad$ ISL3.5P0 = −2

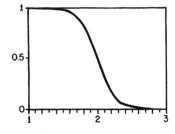

Figure 5.8 Scoring function.

3.6. *Number of Lane Repeats*

Score $\qquad$ IS3.6P0 = SSF (ILTH3.6P0,IB3.6P0,IUTH3.6P0,ISL3.6P0)

Lower Threshold $\qquad$ ILTH3.6P0 = 0

Baseline $\qquad$ IB3.6P0 = 3

Upper Threshold $\qquad$ IUTH3.6P0 = ∞

Slope $\qquad$ ISL3.6P0 = −3

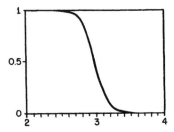

Figure 5.9 Scoring function.

3.7. *Difference Between Fast and Slow*

Score   IS3.7P0 = SSF (ILTH3.7P0,IB3.7P0,IUTH3.7P0,ISL3.7P0)

Lower Threshold   ILTH3.7P0 = 0

Baseline   IB3.7P0 = 2

Upper Threshold   IUTH3.7P0 = 10

Slope   ISL3.7P0 = –3

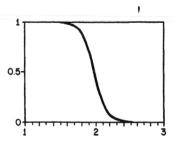

Figure 5.10 Scoring function.

4. *Availability*

Score   IS4P0 = SSF (ILTH4P0,IB4P0,IUTH4P0,ISL4P0)

Lower Threshold   ILTH4P0 = 0

Baseline   IB4P0 = 0.5

Upper Threshold   IUTH4P0 = 1

Slope   ISL4P0 = 2

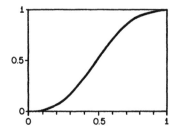

Figure 5.11 Scoring function.

5. *Reliability*

Score   IS5P0 = SSF (ILTH5P0,IB5P0,IUTH5P0,ISL5P0)

Lower Threshold   ILTH5P0 = 0

Baseline   IB5P0 = 0.5

Upper Threshold   IUTH5P0 = 1

Slope   ISL5P0 = 2

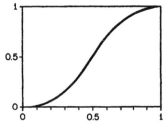

Figure 5.12 Scoring function.

5.3.4.3 Weighting criteria

The following importance values, on a scale from 1 to 10, were assigned to each performance figure of merit. The resultant weight, IWiP0, was computed by summing all the importance values and dividing each entry by this total.

Figure of Merit	Value	IWiP0
1. Average Races per Car	5	0.147059
2. Number of Ties	3	0.088235
3. Happiness	10	0.294118
3.1. Percent Happy Scouts	10	0.238095
3.2. Number Irate Parents	6	0.142857
3.3. Number of Broken Cars	7	0.166667
3.4. Others Touching Scout's Car	4	0.095238
3.5. Number of Repeat Races	6	0.142857
3.6. Number of Lane Repeats	5	0.119048
3.7. Difference Between Fast and Slow	4	0.095238
4. Availability	8	0.235294
5. Reliability	8	0.235294

> *Notice the grouping of the subitems under Happiness. The net result of this is the significant reduction in importance of these factors. The total score that can be achieved by Happiness is 1.0 times the weight. Each item under this heading is weighted so that the category Happiness achieves a score of 1.0 when all those items are at their optimum value. Grouping is necessary to make sense of many related items and can keep them from becoming too important, but its limitations must be recognized.*

5.3.5 Utilization of Resources Requirement

5.3.5.1 Definition of Resource Figures of Merit

The overall Utilization of Resources Figure of Merit is denoted UF0P0 and is computed by

$$UF0P0 = USF1P0 * UW1P0 + USF2P0 * UW2P0 + \ldots + USFnP0 * UWnP0$$

where n is the total number of Utilization of Resources Figures of Merit and

$$USFiP0 = USiP0(IFiP0(FSD)) \text{ for } i = 1 \text{ to } n$$

as will be shown in the next section.

5.3.5.2   Lower, upper, baseline, and scoring parameters
In this section, the following naming convention for variables is used: The initial letter "U" indicates that the name is for a Utilization of Resources Requirement. The terminal P0 indicates that the name involves Problem 0 of the Pinewood Derby.

$UFiP0$ = the ith Utilization of Resources figure of merit.
$UBiP0$ = the baseline value for the ith figure of merit.
$ULTHiP0$ = lower threshold for the ith figure of merit.
$USFiP0$ = score for the ith figure of merit.
$USiP0$ = scoring function for the ith figure of merit.
$USLiP0$ = slope for the ith figure of merit.
$UUTHiP0$ = upper threshold for the ith figure of merit.
$UWiP0$ = weight of the ith figure of merit.
SSF = standard scoring function.

1. *Acquisition Time (in hours)*

   Score $US1P0 = SSF (ULTH1P0,UB1P0,UUTH1P0,USL1P0)$

   Lower Threshold $ULTH1P0 = 0$

   Baseline $UB1P0 = 40$

   Upper Threshold $UUTHiP0 = 400$

   Slope $USL1P0 = -0.05$

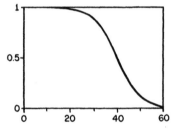

Figure 5.13   Scoring function.

2. *Acquisition Cost (in dollars)*

   Score $US2P0 = SSF (ULTH2P0,ULB2P0,ULSL2P0,UOPT2P0,$
                                              $UUB2P0,UUTH2P0,UUSL2P0)$

   Lower Threshold $ULTH2P0 = -\infty$

   Lower Baseline $ULB2P0 = 0$

   Lower Slope $ULSL2P0 = 0.033$

   Optimum $UOPT2P0 = 50$

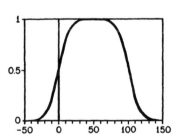

Figure 5.14   Scoring function.

Upper Baseline $UUB2P0 = 100$

Upper Threshold $UUTH2P0 = 300$

Upper Slope $UUSL2P0 = -0.033$

3. *Total Event Time (in hours)*

Score $US3P0 = SSF(ULTH3P0,ULB3P0,ULSL3P0,UOPT3P0,$
$UUB3P0,UUTH3P0,UUSL3P0)$

Lower Threshold $ULTH3P0 = 0$

Lower Baseline $ULB3P0 = 2$

Lower Slope $ULSL3P0 = 0.67$

Optimum $UOPT3P0 = 3.5$

Upper Baseline $UUB3P0 = 4.5$

Upper Threshold $UUTH3P0 = 8$

Slope $UUSL3P0 = -1$

We used a biphasic, hill-shaped scoring function for this figure of merit because we thought the event should last about 3.5 hours. In the months before the race, the scouts in the pack spent about 1000 boy-hours building their cars. For such an investment in time they want an event that lasts a significant amount of time. Anything less than one hour would trivialize their efforts. On the other hand, if the event took more than 5 hours the adults would be exhausted.

4. *Number of Electrical Circuits*

Score   US4P0 = SSF (ULTH4P0,UB4P0,UUTH4P0,USL4P0)

Lower Threshold   ULTH4P0 = 0

Baseline   UB4P0 = 1

Upper Threshold   UUTH4P0 = 6

Slope   USL4P0 = −1

5. *Number of Adults*

Score   US5P0 = SSF (ULTH5P0,UB5P0,UUTH5P0,USL5P0)

Lower Threshold   ULTH5P0 = 1

Baseline   UB5P0 = 5

Upper Threshold   UUTH5P0 = 15

Slope   USL5P0 = −0.25

5.3.5.3 *Weighting criteria*
The following importance values, on a scale from 1 to 10, were assigned to each Utilization of Resources Figure of Merit. Each resultant weight, UWiP0, was computed by summing all the importance values and dividing each entry by this total.

Figure of Merit	Value	UWiP0
1. Acquisition Time	10	0.322581
2. Acquisition Cost	6	0.193548
3. Total Event Time	8	0.258065
4. Number of Electrical Circuits	3	0.096774
5. Number of Adults	2	0.129032

5.3.6 Trade-Off Requirement

The Trade-Off Requirement is computed by the formula

$$TF0P0 = TW1P0 * IFX0P0 + TW2P0 * UFX0P0$$

where TW1P0 is the weight of the Overall I/O Performance Index and TW2P0 is the weight of the overall Utilization of Resources Index. IFX0P0(FSD) indicates the overall score for the feasible I/O Performance Requirement. UFX0P0 indicates the overall score for the feasible U/R Requirement.

For our initial design we will use the following weights:

$$TW1P0 = 0.9$$
$$TW2P0 = 0.1$$

5.3.7 System Test Requirement

5.3.7.1 Test plan

5.3.7.1.1 Explanation of test plan. The test plan will be based on data submitted for simulation before an actual system is developed. Since there is no time or money for an actual system test before deployment, we will base our selection on the results of our simulation using the test trajectories. The test trajectories are based on actual data collected during the 1991 Pinewood Derby.

The system will be acceptable if

1. all requirements from this document are satisfied,
2. the system allows for adverse weather conditions,
3. no more than 1500 square feet are used,
4. the system is completed by the first Sunday in February, and
5. restroom facilities are available for participants.

The system will be in compliance if the upper and lower bounds set for each figure of merit are met.

The figures of merit are measured as described under each test trajectory for each of the tests where appropriate. The results are summed and entered in concept selection data sheets.

These are the product failure modes:

1. Electrical failure (if the final system uses electricity) including
    1.1. total loss of electric power and
    1.2. computer failure (if the final system uses computers).
2. Adverse weather conditions preventing the Derby from being completed.
3. Mistakes in judging race finishes or recording results.
4. Human mistakes in
    4.1. weighing the cars,

    4.2. allowing car modifications after inspection,

    4.3. getting the cars in the correct lanes,

    4.4. resetting the finish line switches (if they are used), and

    4.5. wasting time.

   5. Track imperfections that cause one lane to be faster than another.

The Grand Marshall will determine if any of these product failure modes are entered during the Derby.

    5.3.7.1.2 Test Trajectory 1.   Test Trajectory 1 will determine the system performance through the use of the data for 23 cub scouts from each division. The actual data from the 1991 Pinewood Derby, shown in Exhibit 5.3, will be used as input trajectories in a computer simulation to estimate racing results.

    5.3.7.1.3 Test Trajectory 2.   Test Trajectory 2 will determine if the race judging components are fair. Several cars with similar speeds will be used. Dr. Bahill and Bill Karnavas will be the judges. Forty-six races will be run (in round-robin format with 23 entries), and in each a winner or a tie is declared. Ties will be counted and used as a performance figure of merit.

   5.3.7.2 Input/output performance tests

   1. *Average Races per Car:* This will be calculated by dividing the sum of the number of races for each car by the total number of cars that raced based on Test Trajectories 1 and 2.

   2. *Number of Ties:* The number of ties are observed during the event either visually (human) or automatically (computer sensing device) based on Test Trajectory 2.

   3. *Happiness:* This is a computed measure based on Figures of Merit 5.3.1 through 5.3.7.

     3.1. *Percent Happy Scouts:* This figure of merit is calculated by dividing the number of happy scouts leaving the event by the total number of scouts attending. A happy scout is defined as one that leaves the event looking happy, contented, or pleased. Since this determination is subjective, the final decision will be made by the Race Marshall. It will be based partly on the results of Test Trajectory 1.

     3.2. *Number Irate Parents:* This figure of merit will be determined by the committee volunteers during the event. An irate parent is defined as one that disputes the result of a race or some other judging decision or who makes rude or inappropriate remarks to judges. It will be based on the results of Test Trajectories 1 and 2. Since this is subjective, the final output will be decided by Grand Marshall.

EXHIBIT 5.3

Raw Data from the Webelos Division of the Pack 212 1991 Pinewood Derby

Round	Race	Lane	Letter	Time	Place
1	1	1	A	2.5813	First
1	1	2	B	2.6603	Third
1	1	3	C	2.6200	Second
1	2	1	D	2.5779	First
1	2	2	E	Did not Finish	
1	2	3	F	2.7185	Second
1	3	1	G	2.6301	First
1	3	2	H	2.7010	Second
1	3	3	I	3.3249	Third
1	4	1	J	2.6370	First
1	4	2	K	2.8017	Second
1	4	3	L	2.8209	Third
1	5	1	M	2.9979	Third
1	5	2	N	2.6052	First
1	5	3	O	2.6454	Second
1	6	1	P	2.8248	Third
1	6	2	Q	2.5749	First
1	6	3	R	2.6750	Second
1	7	1	S	2.5837	First
1	7	2	T	2.5898	Second
1	7	3	U	2.6382	Third
1	8	1	V	3.0123	Second
1	8	2	W	2.7434	First
2	1	1	L	2.8310	Second
2	1	2	P	2.9599	Third
2	1	3	F	2.7036	First
2	2	1	B	2.6450	Second
2	2	2	M	3.2100	Third
2	2	3	Q	2.5768	First
2	3	1	H	2.7083	Second
2	3	2	W	2.7709	Third
2	3	3	A	2.5892	First
2	4	1	T	2.5720	First
2	4	2	C	2.6224	Second
2	4	3	V	2.8033	Third
2	5	1	S	2.5739	First
2	5	2	D	2.6139	Second
2	5	3	I	2.9690	Third
2	6	1	E	2.5982	First
2	6	2	N	2.6105	Second

Round	Race	Lane	Letter	Time	Place
2	6	3	G	2.6037	First
2	7	1	J	2.7318	Third
2	7	2	U	2.6614	Second
2	7	3	R	2.6370	Second
2	8	1	K	2.7612	Third
2	8	2	O	2.5880	First
3	1	1	L	2.9174	Second
3	1	2	G	2.6172	First
3	1	3	W	2.7918	Third
3	2	1	N	2.6160	Second
3	2	2	Q	2.5635	First
3	2	3	M	3.0267	Third
3	3	1	J	2.7044	Second
3	3	2	H	2.6813	First
3	3	3	A	2.5632	First
3	4	1	F	2.6709	Second
3	4	2	I	2.9112	Third
3	4	3	R	2.6446	Second
3	5	1	S	2.6641	Third
3	5	2	B	2.6279	First
3	5	3	C	2.5735	First
3	6	1	O	2.6450	Second
3	6	2	P	2.8474	Third
3	6	3	K	2.7667	Third
3	7	1	T	2.6426	Second
3	7	2	D	2.5989	First
3	7	3	E	2.5822	First
3	8	1	U	2.6755	Second
3	8	2	V	2.8769	Third
4	1	1	N	2.6405	First
4	1	2	S	2.6503	Second
4	1	3	P	2.8917	Third
4	2	1	F	2.6738	Second
4	2	2	B	2.6522	First
4	2	3	W	2.7659	Third
4	3	1	Q	2.5961	First
4	3	2	O	2.6072	Second
4	3	3	G	2.6481	Third
4	4	1	J	2.7152	Third
4	4	2	R	2.6936	Second
4	4	3	T	2.6397	First
4	5	1	U	2.6858	First
4	5	2	H	3.1447	Second
4	5	3	V	2.8496	Third
4	6	1	A	2.6222	First
4	6	2	D	2.6275	Second

Round	Race	Lane	Letter	Time	Place
4	6	3	I	2.9596	Third
4	7	1	L	2.9526	Third
4	7	2	C	2.6647	First
4	7	3	E	2.5839	First
4	8	1	M	3.2985	Third
4	8	2	K	2.7837	Second
5	1	1	W	2.7738	Third
5	1	2	C	2.6280	Second
5	1	3	E	2.5887	First
5	2	1	F	2.6618	Second
5	2	2	Q	2.6184	First
5	2	3	J	2.7273	Third
5	3	1	U	2.6384	First
5	3	2	P	2.9492	Third
5	3	3	B	2.6701	Second
5	4	1	I	2.9798	Second
5	4	2	K	2.7707	First
5	4	3	D	2.5896	First
5	5	1	G	2.6432	Second
5	5	2	M	3.0676	Third
5	5	3	T	2.5643	First
5	6	1	A	2.5925	Second
5	6	2	L	2.8194	Third
5	6	3	H	2.8914	Third
5	7	1	R	2.6579	Second
5	7	2	N	2.6097	First
5	7	3	O	2.5995	First
5	8	1	V	2.8129	Third
5	8	2	S	2.6071	Second
6	1	1	K	2.7295	Second
6	1	2	U	2.6894	First
6	1	3	W	2.8001	Third
6	2	1	L	2.9437	Second
6	2	2	V	2.8436	First
6	2	3	M	2.9730	Third
6	3	1	N	2.6021	Second
6	3	2	A	2.5825	First
6	3	3	B	2.6655	Second
6	4	1	D	2.6212	First
6	4	2	J	2.6748	Third
6	4	3	O	2.6255	Second
6	5	1	F	2.7286	Third
6	5	2	T	2.6215	First
6	5	3	C	2.6036	First
6	6	1	G	2.6718	Third
6	6	2	R	2.6418	Second

Round	Race	Lane	Letter	Time	Place
6	6	3	P	2.9009	Third
6	7	1	H	2.7924	Second
6	7	2	E	2.5737	First
6	7	3	Q	2.5610	First
6	8	1	I	2.9521	Third
6	8	2	S	2.5989	Second

3.3. *Number of Broken Cars:* The committee volunteers keep a count of all the broken cars. The final output will be decided by the Grand Marshall.

3.4. *Others Touching Scout's Car:* This is a count of the number of people who touch the scout's cars throughout a race, as observed by the Grand Marshall.

3.5. *Number of Repeat Races:* This is based on the simulation results from Test Trajectory 1.

3.6. *Number of Lane Repeats:* This is based on the simulation results from Test Trajectory 1.

3.7. *Difference Between Fast and Slow:* This is based on the simulation results from Test Trajectory 1. It is the difference between the number of races for the fastest car and the number of races for the slowest car.

4. *Availability:* This is determined through observation by the committee members at the beginning of the Pinewood Derby. If the system works properly initially, then a figure of merit of 1 is recorded; if the system works for most events but fails for some, 0.8 is recorded; if the system barely works at start-up, then 0.2 is recorded; otherwise, 0 is recorded.

5. *Reliability:* This figure of merit will be determined through observation by the committee members throughout the race. If the system at any time shows signs of not properly conducting races or recording races, it shall be deemed unreliable and a score of 0.8 is recorded. If the system fails often, a score of 0.2 is recorded. If the system fails to work at least half the time, a score of 0 is recorded. If the system always works, a score of 1.0 is recorded.

5.3.7.3   Utilization of resources tests

1. *Acquisition Time:*   This figure of merit represents the number of hours it takes to complete the project, as observed by Dr. Bahill. The minimum value is 0 and the maximum is 400.

2. *Acquisition Cost:*   This figure of merit is an approximation by Dr. Bahill of the cost of designing and implementing the system.

3. *Total Event Time:*   The total event time will be calculated by subtracting the start time from the end time.

4. *Number of Electrical Circuits:*   The system designers will estimate the total number of 120 VAC, 15 A, circuits the system will require.

5. *Number of Adults:*   Dr. Bahill will count the number of adults needed.

5.3.8   Rationale for operational need

Data for this document were provided by Dr. Bahill, Bill Karnavas, and the Cub Master.

Harry Williams has been the Cub Master for Pack 212 for the past decade. Dr. Bahill and Bill Chapman interviewed him at his home on September 17, 1990. The items listed below summarize his comments.

- The main purpose of the derby is to entertain the scouts; the competition is what makes it fun. We feel that the cars should be raced at least three times to make it worth the effort of creating the vehicle.

- We like to know the results of division races during the races. If the format is too technical, we can't understand it. Results can be posted by computer display or handwritten notes on a corkboard.

- The round-robin format was used in 1989 and 1990. It is fairer and the scouts get to race their cars more often, though we better understand the double-elimination tournament, which is the technique most packs use. The problem with a double-elimination tournament is that a scout might get to race only twice and he may not race in what he thinks is the best lane.

- The main parental complaint was about weighing the cars. A car's weight limit is 5 ounces, but some weigh in slightly over 5 ounces. The parents say that the pack's scale must be wrong, since they already weighed below the limit at home.

- There is a perception of unfairness in judging when human judges are used, but this perception decreases a lot with computer timing.

- We can get as many adults as we need to manage the races. Last year we used eight.

- We had only 600 square feet of space to run the races last year; crowd control was a problem. More room would help the scouts to see the races and prevent confusion and damage to their cars. We need parking space for at least 30 automobiles.

- Bleachers would enable everyone to see the races.

- An upper limit of six hours for the entire derby is reasonable.

- We think 50 linear feet of storage space is optimal for cars between races.

- A major disappointment for a scout is when his car does not make it to the bottom of the track because of design flaws. Other boys laugh and his feelings get hurt. We don't know what can be done to avoid this.

5.4 Document 4: System Requirements Validation

In the System Requirements Validation Document we

(1) examine the mathematical description of the requirements presented in Document 3 to check for consistency,

(2) demonstrate that a real world solution can be built, and

(3) show that a real world solution can be tested to prove that it satisfies the requirements.

If the client has requested a perpetual motion machine or a system that reduces entropy, this is the time to stop the project and save money.

5.4.1 Input/output and functional design

After examining the required inputs and outputs for the Pinewood Derby, it is obvious to us that all of the requirements had been satisfied (although not optimized) in prior years. All of the information needed for this examination was easily obtained. Therefore, we are satisfied that the system's inputs and outputs are feasible.

5.4.2 *Technology for the buildable system design*

An examination of the Technology Requirement shows nothing that inhibits the functioning of the system. Derbies in the past easily fit within these requirements.

5.4.3 *Input/Output Performance Requirement*

All the requirements in this category have been satisfied in past derbys. The most restrictive is the limiting of the number of ties to an upper threshold of five. That number is based on 23 entries in the event. Two closely matched cars may present a problem if the judging resolution is poor. Available technology includes computerized monitoring of the finish line. This will provide a resolution of 0.0001 second, which is accurate enough to prevent ties. Therefore, this requirement can be met.

5.4.4 *Utilization of Resources Requirement*

The requirements of this section also have been met in prior derbys. The most restrictive requirement is the upper limit of $300 on acquisition cost. The timing mechanism needed to ensure that only few ties occur could be expensive, but we have found that using a borrowed computer for processing and purchasing switches for installation at the bottom of the track can be done for less than $300.

5.4.5 *Test Requirement*

No problems are foreseen in meeting the acceptability, compliance, or observability requirements of this section.

5.5 Document 5: Concept Exploration

The Concept Exploration Document is used to study several different system designs via approximation, simulation, or prototypes, or via a combination of these techniques. The best design alternative is suggested by the data. This document will be rewritten many times as more information becomes available.

5.5.1 System design concepts

5.5.1.1 System Design Concept 1
System Design Concept 1 specifies a single-elimination tournament. The winner of each race will advance to the next race. One loss will eliminate a participant from the tournament.

5.5.1.2 System Design Concept 2
System Design Concept 2 specifies a double-elimination tournament. Each participant is allowed one loss without elimination, that is, one finish short of first place. First place finishers go on to race only first place finishers; those with one loss race others with one loss. The overall first place winner is the only participant not to be eliminated. The overall second place winner is the car that lost its last race against the first place finisher, and the third place winner is the second to last car to lose two races.

5.5.1.3 System Design Concept 3
System Design Concept 3 specifies a round-robin tournament with mean-time scoring to determine overall winners. The round robin is scheduled so that every contestant races at least once in each lane and against as diverse a number of entries as possible. This should help prevent any problems with lane bias, and it will add to the number of races for each contestant. Suitable schedules for such round-robin tournaments are given in Section 5.8 of this chapter. In the mean-time scoring system, the race times of each contestant for all his races are averaged; the participant with the lowest mean time is the first place finisher, the second-lowest is the second place finisher, and the third-lowest is the third place finisher.

The median time might be better than the mean time because sometimes a race can be a disaster, with the car falling off the track or a very slow finish of three to four times the car's average time. This kind of poor finish very heavily influences the average time, putting such a contestant essentially out of the running. The median is also easier to calculate than the mean.

5.5.1.4 *System Design Concept 4*

System Design Concept 4 specifies a round-robin tournament with best-time scoring to determine overall winners. The round robin will be scheduled so that every contestant races at least once in each lane and against as diverse a number of entries as possible. This should help prevent any problems with lane bias, and it will add to the number of races for each contestant. In the best-time scoring system, the fastest race time of each contestant in each race is recorded. The lowest time is the first place finisher, the second-lowest is the second place finsher, and the third-lowest is the third place finisher.

5.5.1.5 *System Design Concept 5*

System Design Concept 5 specifies a round-robin tournament with point assignment scoring to determine overall winners. The round robin will be scheduled so that every contestant races at least once in each lane and against as diverse a number of entries as possible. This should help prevent any problems with lane bias, and it will add to the number of races for each contestant. In the point assignment scoring system, a first place finish in a race will be assigned three points; a second place finish, two points; and a third place finish, one point. The contestant with the highest total score is the overall first place finisher, the second highest score is the second place finisher, and the third highest score is the third place finisher.

The advantage of Concept 5 over Concepts 3 and 4 is that exact times from each race are not necessary, only a determination of who came in first, second, and third. Thus, this concept can be easily implemented using low temporal resolution judging, such as that provided by humans.

5.5.1.6 *System Design Concept 6*

System Design Concept 6 specifies that human judges will determine race results. The resolving ability of a human judge is approximately 0.01 second (or 1 inch).

5.5.1.7　System Design Concept 7

System Design Concept 7 specifies that electronic circuits will determine race results. A switch in each lane is triggered as a car passes the finish line. The resolution of the electronic judging system is 0.0001 second.

> *The seven concepts above are not independent concepts. They provide alternatives for two independent subproblems—five alternatives for the racing format and two alternatives for judging technique. One alternative must be selected from each category. In general, some system designs will list only one subproblem and others will list many.*

5.5.2　Figures of merit

The figures of merit are calculated using the test plan described in Document 3 and based on the systems described in Documents 6 and 7. The values obtained for these figures of merit are entered here, then the scores are computed using the standard scoring functions defined in Document 3. The formulas

$$IF0P0(FSDi) = IW1P0 * ISF1P0 + \ldots + IWmP0 * ISFmP0$$

$$UF0P0(FSDi) = UW1P0 * USF1P0 + \ldots + UWnP0 * USFnP0$$

are used to compute the overall figures of merit for each design, where m is the number of I/O Performance Figures of Merit and n is the number of resource figures of merit, and

$$ISF1P0 = IS1P0(IFX1P0(FSDi))$$

$$USF1P0 = US1P0(UFX1P0(FSDi))$$

where i is the concept design number.

The tables on the following pages show the estimates given for the figures of merit. The column titled IFXiP0 (where i is the figure of merit number) is the figure of merit measured per the test plan. The column labeled ISFiP0 is the calculated score after entering the figure of merit into the standard scoring function defined in Document 3. The column IWiP0 is the weight factor given in Document 3 for the respective figure of merit. The overall scores, IF0P0 and UF0P0, are determined from the weights and scores.

When there are sub-requirements, a cascade process is followed. First, the value of the figure of merit is obtained; for example, in the table in Section 5.5.2.1.1 we expected five irate parents, so the value 5 is entered in the intersection of the row labeled "3.2. Number Irate Parents" and the IFXiP0 column. This value is next processed through its scoring function, in this example yielding a score of 0.0. This score is multiplied by its weight, 0.142857. The weighted scores for the seven sub-requirements 3.1 to 3.7 are calculated and then added together. This total is the figure of merit value (the IFXiP0 column) for the requirement "3. Happiness" (0.398 in this example). Now we perform the second step of passing this value through its scoring function to get its score of 0.306; this score is then multiplied by its weight, 0.294118. Finally, the weighted scores of all five requirements are summed to give the overall performance figure of merit of 0.656 for the approximation data for this concept.

Three different methods for determining the figures of merit are given: approximation, simulation, and prototype. These methods reflect the different types of data available for determining figures of merit throughout the initial design. Approximation values are based on estimates made by the systems engineer based on experience and historical data. Simulation data are obtained using models built to simulate the prototype. Prototype data are calculated from previous derbies.

In the tables below, the figure of merit Number of Ties is treated differently for Concepts 1 to 5 than for Concepts 6 and 7. The number of ties is a function of the judging technique, not of race format. Therefore, values of 0 were entered for the figure of merit for Concepts 1 to 5, whereas actual numbers were used for Concepts 6 and 7. Given this philosophy, perhaps it would have been better to call the figure of merit "Percentage of Races Called a Tie."

5.5.2.1 Figures of merit for Concept 1
Concept 1 specifies a single-elimination tournament. Tables for the approximation and simulation methods follow.

5.5.2.1.1 Approximation figures of merit for Concept 1

I/O FIGURES OF MERIT

REQUIREMENTS	IFXiP0(FSD1)	ISFiP0(FSD1)	IWiP0
1 Average Races per Car	2	0.051	0.147059
2 Number of Ties	0	1	0.088235
3 Happiness	0.398	0.306	0.294118
3.1 Percent Happy Scouts	50	0	0.238095
3.2 Number Irate Parents	5	0	0.142857
3.3 Number of Broken Cars .	1.2	0.310	0.166667
3.4 Others Touching Scout's Car	1	0.889	0.095238
3.5 Number of Repeat Races	0	1	0.142857
3.6 Number of Lane Repeats	1	1	0.119048
3.7 Difference Between Fast and Slow	5	0	0.095238
4 Availability	1	1	0.235294
5 Reliability	1	1	0.235294

IF0P0(FSD1) = 0.656

U/R FIGURES OF MERIT

REQUIREMENTS	UFXiP0(FSD1)	USFiP0(FSD1)	UWiP0
1 Acquisition Time	10	0.97	0.322581
2 Acquisition Cost	10	0.79	0.193548
3 Total Event Time	2	0.5	0.258065
4 Number of Electrical Circuits	0	1	0.096774
5 Number of Adults	4	0.732	0.129032

UF0P0(FSD1) = 0.786

5.5.2.1.2 Simulation figures of merit for Concept 1

I/O FIGURES OF MERIT

REQUIREMENTS	IFXiP0(FSD1)	ISFiP0(FSD1)	IWiP0
1 Average Races per Car	1.4	0.01	0.147059
2 Number of Ties	0	1	0.088235
3 Happiness	0.37	0.260	0.294118
3.1 Percent Happy Scouts	50	0	0.238095
3.2 Number Irate Parents	5	0	0.142857
3.3 Number of Broken Cars	1	0.5	0.166667
3.4 Others Touching Scout's Car	1	0.889	0.095238
3.5 Number of Repeat Races	0	1	0.142857
3.6 Number of Lane Repeats	3	0.5	0.119048
3.7 Difference Between Fast and Slow	4	0	0.095238
4 Availability	1	1	0.235294
5 Reliability	1	1	0.235294

IF0P0(FSD1) = 0.637

U/R FIGURES OF MERIT

REQUIREMENTS	UFXiP0(FSD1)	USFiP0(FSD1)	UWiP0
1 Acquisition Time	10	0.97	0.322581
2 Acquisition Cost	10	0.79	0.193548
3 Total Event Time	2	0.5	0.258065
4 Number of Electrical Circuits	0	1	0.096774
5 Number of Adults	4	0.732	0.129032

UF0P0(FSD1) = 0.786

5.5.2.2   Figures of merit for Concept 2

Concept 2 specifies a double-elimination tournament. Tables for the approximation, simulation, and prototype methods follow.

5.5.2.2.1   Approximation figures of merit for Concept 2

I/O FIGURES OF MERIT

REQUIREMENTS	IFXiP0(FSD2)	ISFiP0(FSD2)	IWiP0
1 Average Races per Car	3	0.206	0.147059
2 Number of Ties	0	1	0.088235
3 Happiness	0.377	0.271	0.294118
3.1 Percent Happy Scouts	85	0.119	0.238095
3.2 Number Irate Parents	4	0	0.142857
3.3 Number of Broken Cars	2	0.015	0.166667
3.4 Others Touching Scout's Car	1	0.889	0.095238
3.5 Number of Repeat Races	1	1	0.142857
3.6 Number of Lane Repeats	2	1	0.119048
3.7 Difference Between Fast and Slow	4	0	0.095238
4 Availability	1	1	0.235294
5 Reliability	1	1	0.235294

IF0P0(FSD2) = 0.669

U/R FIGURES OF MERIT

REQUIREMENTS	UFXiP0(FSD2)	USFiP0(FSD2)	UWiP0
1 Acquisition Time	10	0.97	0.322581
2 Acquisition Cost	20	0.937	0.193548
3 Total Event Time	5	0.119	0.258065
4 Number of Electrical Circuits ·	0	1	0.096774
5 Number of Adults	6	0.269	0.129032

UF0P0(FSD2) = 0.656

5.5.2.2.2 Simulation figures of merit for Concept 2

I/O FIGURES OF MERIT

REQUIREMENTS	IFX*i*P0(FSD2)	ISF*i*P0(FSD2)	IW*i*P0
1 Average Races per Car	3.7	0.401	0.147059
2 Number of Ties	0	1	0.088235
3 Happiness	0.241	0.103	0.294118
3.1 Percent Happy Scouts	90	0.5	0.238095
3.2 Number Irate Parents	4	0	0.142857
3.3 Number of Broken Cars	2	0.015	0.166667
3.4 Others Touching Scout's Car	2	0.5	0.095238
3.5 Number of Repeat Races	2	0.5	0.142857
3.6 Number of Lane Repeats	8	0	0.119048
3.7 Difference Between Fast and Slow	3	0	0.095238
4 Availability	1	1	0.235294
5 Reliability	1	1	0.235294

IF0P0(FSD2) = 0.648

U/R FIGURES OF MERIT

REQUIREMENTS	UFX*i*P0(FSD2)	USF*i*P0(FSD2)	UW*i*P0
1 Acquisition Time	10	0.970	0.322581
2 Acquisition Cost	·20	0.937	0.193548
3 Total Event Time	5	0.119	0.258065
4 Number of Electrical Circuits	0	1	0.096774
5 Number of Adults	6	0.269	0.129032

UF0P0(FSD2) = 0.656

5.5.2.2.3   Prototype figures of merit for Concept 2 (from the 1988 Derby)

I/O FIGURES OF MERIT

REQUIREMENTS	IFXiP0(FSD2)	ISFiP0(FSD2)	IWiP0
1 Average Races per Car	3.7	0.401	0.147059
2 Number of Ties	0	1	0.088235
3 Happiness	0.138	0.036	0.294118
3.1 Percent Happy Scouts	80	0.015	0.238095
3.2 Number Irate Parents	7	0	0.142857
3.3 Number of Broken Cars	2	0.015	0.166667
3.4 Others Touching Scout's Car	5	0.002	0.095238
3.5 Number of Repeat Races	2	0.5	0.142857
3.6 Number of Lane Repeats	8	0.5	0.119048
3.7 Difference Between Fast and Slow	3	0	0.095238
4 Availability	1	1	0.235294
5 Reliability	1	1	0.235294

IF0P0(FSD2) = 0.628

U/R FIGURES OF MERIT

REQUIREMENTS	UFXiP0(FSD2)	USFiP0(FSD2)	UWiP0
1 Acquisition Time	15	0.937	0.322581
2 Acquisition Cost	35	0.994	0.193548
3 Total Event Time	5	0.119	0.258065
4 Number of Electrical Circuits	0	1	0.096774
5 Number of Adults	6	0.269	0.129032

UF0P0(FSD2) = 0.842

5.5.2.3 Figures of merit for Concept 3

Concept 3 specifies a round-robin tournament with mean-time scoring. Tables for the approximation and simulation methods follow.

5.5.2.3.1 Approximation figures of merit for Concept 3

I/O FIGURES OF MERIT

REQUIREMENTS	IFXiP0(FSD3)	ISFiP0(FSD3)	IWiP0
1 Average Races per Car	6	0.935	0.147059
2 Number of Ties	0	1	0.088235
3 Happiness	0.514	0.528	0.294118
3.1 Percent Happy Scouts	95	0.889	0.238095
3.2 Number Irate Parents	2	0.018	0.142857
3.3 Number of Broken Cars	2	0.015	0.166667
3.4 Others Touching Scout's Car	3	0.118	0.095238
3.5 Number of Repeat Races	2	0.5	0.142857
3.6 Number of Lane Repeats	0	1	0.119048
3.7 Difference Between Fast and Slow	0	1	0.095238
4 Availability	1	1	0.235294
5 Reliability	1	1	0.235294

IF0P0(FSD3) = 0.852

U/R FIGURES OF MERIT

REQUIREMENTS	UFXiP0(FSD3)	USFiP0(FSD3)	UWiP0
1 Acquisition Time	2	0.998	0.322581
2 Acquisition Cost	70	0.986	0.193548
3 Total Event Time	6	0.002	0.258065
4 Number of Electrical Circuits	1	0.5	0.096774
5 Number of Adults	7	0.118	0.129032

UF0P0(FSD3) = 0.577

5.5.2.3.2 Simulation figures of merit for Concept 3

I/O FIGURES OF MERIT

REQUIREMENTS	IFXiP0(FSD3)	ISFiP0(FSD3)	IWiP0
1 Average Races per Car	6	0.935	0.147059
2 Number of Ties	0	0	0.088235
3 Happiness	0.514	0.528	0.294118
3.1 Percent Happy Scouts	95	0.889	0.238095
3.2 Number Irate Parents	2	0.018	0.142857
3.3 Number of Broken Cars	2	0.015	0.166667
3.4 Others Touching Scout's Car	3	0.118	0.095238
3.5 Number of Repeat Races	2	0.5	0.142857
3.6 Number of Lane Repeats	0	1	0.119048
3.7 Difference Between Fast and Slow	0	1	0.095238
4 Availability	1	1	0.235294
5 Reliability	1	1	0.235294

IF0P0(FSD3) = 0.852

U/R FIGURES OF MERIT

REQUIREMENTS	UFXiP0(FSD3)	USFiP0(FSD3)	UWiP0
1 Acquisition Time	2	0.998	0.322581
2 Acquisition Cost	70	0.986	0.193548
3 Total Event Time	6	0.002	0.258065
4 Number of Electrical Circuits	1	0.5	0.096774
5 Number of Adults	7	0.118	0.129032

UF0P0(FSD3) = 0.577

5.5.2.4 Figures of merit for Concept 4

Concept 4 specifies a round-robin tournament with best-time scoring. Tables for the approximation, simulation, and prototype methods follow.

5.5.2.4.1 Approximation figures of merit for Concept 4

I/O FIGURES OF MERIT

REQUIREMENTS	IFXiP0(FSD4)	ISFiP0(FSD4)	IWiP0
1 Average Races per Car	6	0.935	0.147059
2 Number of Ties	0	1	0.088235
3 Happiness	0.535	0.570	0.294118
3.1 Percent Happy Scouts	98	0.979	0.238095
3.2 Number Irate Parents	2	0.018	0.142857
3.3 Number of Broken Cars	2	0.015	0.166667
3.4 Others Touching Scout's Car	3	0.118	0.095238
3.5 Number of Repeat Races	2	0.5	0.142857
3.6 Number of Lane Repeats	0	1	0.119048
3.7 Difference Between Fast and Slow	0	1	0.095238
4 Availability	1	1	0.235294
5 Reliability	1	1	0.235294

IF0P0(FSD4) = 0.864

U/R FIGURES OF MERIT

REQUIREMENTS	UFXiP0(FSD4)	USFiP0(FSD4)	UWiP0
1 Acquisition Time	2	0.998	0.322581
2 Acquisition Cost	70	0.986	0.193548
3 Total Event Time	6	0.002	0.258065
4 Number of Electrical Circuits	1	0.5	0.096774
5 Number of Adults	7	0.118	0.129032

UF0P0(FSD4) = 0.577

5.5.2.4.2 Simulation figures of merit for Concept 4

I/O FIGURES OF MERIT

REQUIREMENTS	IFX*i*P0(FSD4)	ISF*i*P0(FSD4)	IW*i*P0
1 Average Races per Car	6	0.935	0.147059
2 Number of Ties	0	0	0.088235
3 Happiness	0.535	0.570	0.294118
3.1 Percent Happy Scouts	98	0.979	0.238095
3.2 Number Irate Parents	2	0.018	0.142857
3.3 Number of Broken Cars	2	0.015	0.166667
3.4 Others Touching Scout's Car	3	0.118	0.095238
3.5 Number of Repeat Races	2	0.5	0.142857
3.6 Number of Lane Repeats	0	1	0.119048
3.7 Difference Between Fast and Slow	0	1	0.095238
4 Availability	1	1	0.235294
5 Reliability	1	1	0.235294

IF0P0(FSD4) = 0.864

U/R FIGURES OF MERIT

REQUIREMENTS	UFX*i*P0(FSD4)	USF*i*P0(FSD4)	UW*i*P0
1 Acquisition Time	2	0.998	0.322581
2 Acquisition Cost	70	0.986	0.193548
3 Total Event Time	6	0.002	0.258065
4 Number of Electrical Circuits	1	0.5	0.096774
5 Number of Adults	7	0.118	0.129032

UF0P0(FSD4) = 0.577

5.5.2.4.3 Prototype figures of merit for Concept 4 (from the 1991 Derby)

I/O FIGURES OF MERIT

REQUIREMENTS	IFX*i*P0(FSD4)	ISF*i*P0(FSD4)	IW*i*P0
1 Average Races per Car	6	0.935	0.147059
2 Number of Ties	0	1	0.088235
3 Happiness	0.680	0.812	0.294118
3.1 Percent Happy Scouts	96	0.970	0.238095
3.2 Number Irate Parents	0	0.018	0.142857
3.3 Number of Broken Cars	0	0.015	0.166667
3.4 Others Touching Scout's Car	2	0.889	0.095238
3.5 Number of Repeat Races	0	1	0.142857
3.6 Number of Lane Repeats	0	1	0.119048
3.7 Difference Between Fast and Slow	0	1	0.095238
4 Availability	1	1	0.235294
5 Reliability	1	1	0.235294

IF0P0(FSD4) = 0.847

U/R FIGURES OF MERIT

REQUIREMENTS	UFX*i*P0(FSD4)	USF*i*P0(FSD4)	UW*i*P0
1 Acquisition Time	2	0.998	0.322581
2 Acquisition Cost	100	0.986	0.193548
3 Total Event Time	· 3.5	0.002	0.258065
4 Number of Electrical Circuits	2	0.5	0.096774
5 Number of Adults	7	0.118	0.129032

UF0P0(FSD4) = 0.577

5.5.2.5 Figures of merit for Concept 5

Concept 5 specifies a round-robin tournament with point-assignment scoring. Tables for the approximation, simulation, and prototype methods follow.

5.5.2.5.1 Approximation figures of merit for Concept 5

I/O FIGURES OF MERIT

REQUIREMENTS	IFXiP0(FSD5)	ISFiP0(FSD5)	IWiP0
1 Average Races per Car	6	0.935	0.147059
2 Number of Ties	0	1	0.088235
3 Happiness	0.421	0.347	0.294118
3.1 Percent Happy Scouts	90	0.5	0.238095
3.2 Number Irate Parents	2	0.018	0.142857
3.3 Number of Broken Cars	2	0.015	0.166667
3.4 Others Touching Scout's Car	3	0.118	0.095238
3.5 Number of Repeat Races	2	0.5	0.142857
3.6 Number of Lane Repeats	0	1	0.119048
3.7 Difference Between Fast and Slow	0	1	0.095238
4 Availability	1	1	0.235294
5 Reliability	1	1	0.235294

IF0P0(FSD5) = 0.798

U/R FIGURES OF MERIT

REQUIREMENTS	UFXiP0(FSD5)	USFiP0(FSD5)	UWiP0
1 Acquisition Time	2	0.998	0.322581
2 Acquisition Cost	70	0.986	0.193548
3 Total Event Time	6	0.002	0.258065
4 Number of Electrical Circuits .	1	0.5	0.096774
5 Number of Adults	7	0.118	0.129032

UF0P0(FSD5) = 0.577

5.5.2.5.2 Simulation figures of merit for Concept 5

I/O FIGURES OF MERIT

REQUIREMENTS	IFXiP0(FSD5)	ISFiP0(FSD5)	IWiP0
1 Average Races per Car	6	0.935	0.147059
2 Number of Ties	0	1	0.088235
3 Happiness	0.467	0.434	0.294118
3.1 Percent Happy Scouts	92	0.691	0.238095
3.2 Number Irate Parents	2	0.018	0.142857
3.3 Number of Broken Cars	2	0.015	0.166667
3.4 Others Touching Scout's Car	3	0.118	0.095238
3.5 Number of Repeat Races	2	0.5	0.142857
3.6 Number of Lane Repeats	0	1	0.119048
3.7 Difference Between Fast and Slow	0	1	0.095238
4 Availability	1	1	0.235294
5 Reliability	1	1	0.235294

IF0P0(FSD5) = 0.824

U/R FIGURES OF MERIT

REQUIREMENTS	UFXiP0(FSD5)	USFiP0(FSD5)	UWiP0
1 Acquisition Time	2	0.998	0.322581
2 Acquisition Cost	70	0.986	0.193548
3 Total Event Time	6	0.002	0.258065
4 Number of Electrical Circuits	1	0.5	0.096774
5 Number of Adults	7	0.118	0.129032

UF0P0(FSD5) = 0.577

5.5.2.5.3   Prototype figures of merit for Concept 5
(from the 1990 derby)

I/O FIGURES OF MERIT

REQUIREMENTS	IFX*i*P0(FSD5)	ISF*i*P0(FSD5)	IW*i*P0
1 Average Races per Car	6	0.935	0.147059
2 Number of Ties	0	1	0.088235
3 Happiness	0.666	0.434	0.294118
3.1 Percent Happy Scouts	90	0.5	0.238095
3.2 Number Irate Parents	1	0.5	0.142857
3.3 Number of Broken Cars	0	1	0.166667
3.4 Others Touching Scout's Car	3	0.118	0.095238
3.5 Number of Repeat Races	1	1	0.142857
3.6 Number of Lane Repeats	4	0.5	0.119048
3.7 Difference Between Fast and Slow	0	1	0.095238
4 Availability	1	1	0.235294
5 Reliability	0.80	.92	0.235294

IF0P0(FSD5) = 0.804

U/R FIGURES OF MERIT

REQUIREMENTS	UFX*i*P0(FSD5)	USF*i*P0(FSD5)	UW*i*P0
1 Acquisition Time	2	0.998	0.322581
2 Acquisition Cost	250	0	0.193548
3 Total Event Time	4	0.889	0.258065
4 Number of Electrical Circuits	1	0.5	0.096774
5 Number of Adults	7	0.118	0.129032

UF0P0(FSD5) = 0.615

For the tables for Concepts 6 and 7, we inserted zeros for the Average Races per Car figure of merit because this figure of merit had no direct relationship with the judging technique. Perhaps we should have also done this for other figures of merit, such as Number of Broken Cars and Number of Lane Repeats. We should have done a better job in defining the figures of merit, identifying some for choosing the racing format and others for selecting the judging technique.

5.5.2.6 Figures of merit for Concept 6

Concept 6 specifies the use of human judges. Tables for the approximation, simulation, and prototype methods follow.

5.5.2.6.1 Approximation figures of merit for Concept 6

I/O FIGURES OF MERIT

REQUIREMENTS	IFXiP0(FSD6)	ISFiP0(FSD6)	IWiP0
1 Average Races per Car	0	0	0.147059
2 Number of Ties	3	0	0.088235
3 Happiness	0.472	0.444	0.294118
3.1 Percent Happy Scouts	90	0.5	0.238095
3.2 Number Irate Parents	10	0	0.142857
3.3 Number of Broken Cars	1.5	0.118	0.166667
3.4 Others Touching Scout's Car	2	0.5	0.095238
3.5 Number of Repeat Races	2	0.5	0.142857
3.6 Number of Lane Repeats	0	1	0.119048
3.7 Difference Between Fast and Slow	0	1	0.095238
4 Availability	1	1	0.235294
5 Reliability	1	1	0.235294

IF0P0(FSD6) = 0.601

U/R FIGURES OF MERIT

REQUIREMENTS	UFXiP0(FSD6)	USFiP0(FSD6)	UWiP0
1 Acquisition Time	10	0.97	0.322581
2 Acquisition Cost	10	0.79	0.193548
3 Total Event Time	4	0.889	0.258065
4 Number of Electrical Circuits	0	1	0.096774
5 Number of Adults	8	0.046	0.129032

UF0P0(FSD6) = 0.798

5.5.2.6.2   Simulation figures of merit for Concept 6

I/O FIGURES OF MERIT

REQUIREMENTS	IFX*i*P0(FSD6)	ISF*i*P0(FSD6)	IW*i*P0
1 Average Races per Car	0	0	0.147059
2 Number of Ties	5	0	0.088235
3 Happiness	0.326	0.196	0.294118
3.1 Percent Happy Scouts	85	0.119	0.238095
3.2 Number Irate Parents	10	0	0.142857
3.3 Number of Broken Cars	1	0.5	0.166667
3.4 Others Touching Scout's Car	5	0.002	0.095238
3.5 Number of Repeat Races	5	0	0.142857
3.6 Number of Lane Repeats	0	1	0.119048
3.7 Difference Between Fast and Slow	0	1	0.095238
4 Availability	1	1	0.235294
5 Reliability	1	1	0.235294

IF0P0(FSD6) = 0.528

U/R FIGURES OF MERIT

REQUIREMENTS	UFX*i*P0(FSD6)	USF*i*P0(FSD6)	UW*i*P0
1 Acquisition Time	10	0.970	0.322581
2 Acquisition Cost	10	0.79	0.193548
3 Total Event Time	4.2	0.771	0.258065
4 Number of Electrical Circuits	0	1	0.096774
5 Number of Adults	8	0.046	0.129032

UF0P0(FSD6) = 0.767

5.5.2.6.3 Prototype figures of merit for Concept 6
(from the 1989 derby)

I/O FIGURES OF MERIT

REQUIREMENTS	IFXiP0(FSD6)	ISFiP0(FSD6)	IWiP0
1 Average Races per Car	0	0	0.147059
2 Number of Ties	12	0	0.088235
3 Happiness	0.36	0.24	0.294118
3.1 Percent Happy Scouts	85	0.119	0.238095
3.2 Number Irate Parents	6	0	0.142857
3.3 Number of Broken Cars	1	0.5	0.166667
3.4 Others Touching Scout's Car	4	0.022	0.095238
3.5 Number of Repeat Races	2	0	0.142857
3.6 Number of Lane Repeats	0	1	0.119048
3.7 Difference Between Fast and Slow	0	1	0.095238
4 Availability	1	1	0.235294
5 Reliability	1	1	0.235294

IF0P0(FSD6) = 0.541

U/R FIGURES OF MERIT

REQUIREMENTS	UFXiP0(FSD6)	USFiP0(FSD6)	UWiP0
1 Acquisition Time	10	0.970	0.322581
2 Acquisition Cost	100	0.5	0.193548
3 Total Event Time	5	0.889	0.258065
4 Number of Electrical Circuits	0	1	0.096774
5 Number of Adults	8	0.046	0.129032

UF0P0(FSD6) = 0.761

5.5.2.7 Figures of merit for Concept 7

Concept 7 specifies the use of electronic judging. Tables for the approximation, simulation, and prototype methods follow.

5.5.2.7.1 Approximation figures of merit for Concept 7

I/O FIGURES OF MERIT

REQUIREMENTS	IFX*i*P0(FSD7)	ISF*i*P0(FSD7)	IW*i*P0
1 Average Races per Car	0	0	0.147059
2 Number of Ties	1	0.018	0.088235
3 Happiness	0.721	0.86	0.294118
3.1 Percent Happy Scouts	98	0.979	0.238095
3.2 Number Irate Parents	1	0.5	0.142857
3.3 Number of Broken Cars	1	0.5	0.166667
3.4 Others Touching Scout's Car	2	0.5	0.095238
3.5 Number of Repeat Races	2	0.5	0.142857
3.6 Number of Lane Repeats	0	1	0.119048
3.7 Difference Between Fast and Slow	0	1	0.095238
4 Availability	1	1	0.235294
5 Reliability	1	1	0.235294

IF0P0(FSD7) = 0.725

U/R FIGURES OF MERIT

REQUIREMENTS	UFX*i*P0(FSD7)	USF*i*P0(FSD7)	UW*i*P0
1 Acquisition Time	2	0.998	0.322581
2 Acquisition Cost	100	0.5	0.193548
3 Total Event Time	4	0.889	0.258065
4 Number of Electrical Circuits	2	0.018	0.096774
5 Number of Adults	6	0.269	0.129032

UF0P0(FSD7) = 0.685

5.5.2.7.2 Simulation figures of merit for Concept 7

I/O FIGURES OF MERIT

REQUIREMENTS	IFXiP0(FSD7)	ISFiP0(FSD7)	IWiP0
1 Average Races per Car	0	0	0.147059
2 Number of Ties	0	1	0.088235
3 Happiness	0.793	0.924	0.294118
3.1 Percent Happy Scouts	98	0.979	0.238095
3.2 Number Irate Parents	1	0.5	0.142857
3.3 Number of Broken Cars	1	0.5	0.166667
3.4 Others Touching Scout's Car	2	0.5	0.095238
3.5 Number of Repeat Races	0	1	0.142857
3.6 Number of Lane Repeats	0	1	0.119048
3.7 Difference Between Fast and Slow	0	1	0.095238
4 Availability	1	1	0.235294
5 Reliability	1	1	0.235294

IF0P0(FSD7) = 0.831

U/R FIGURES OF MERIT

REQUIREMENTS	UFXiP0(FSD7)	USFiP0(FSD7)	UWiP0
1 Acquisition Time	2	0.998	0.322581
2 Acquisition Cost	100	0.5	0.193548
3 Total Event Time	4	0.889	0.258065
4 Number of Electrical Circuits	2	0.018	0.096774
5 Number of Adults	6	0.269	0.129032

UF0P0(FSD7) = 0.586

5.5.2.7.3   Prototype figures of merit for Concept 7 (from averaging the 1990 and 1991 Derbies)

I/O FIGURES OF MERIT

REQUIREMENTS	IFX*i*P0(FSD7)	ISF*i*P0(FSD7)	IW*i*P0
1 Average Races per Car	0	0	0.147059
2 Number of Ties	0	1	0.088235
3 Happiness	0.752	0.85	0.294118
3.1 Percent Happy Scouts	93	0.6	0.238095
3.2 Number Irate Parents	0.5	0.88	0.142857
3.3 Number of Broken Cars	0	1	0.166667
3.4 Others Touching Scout's Car	2.5	0.38	0.095238
3.5 Number of Repeat Races	0.5	0.88	0.142857
3.6 Number of Lane Repeats	2	0.5	0.119048
3.7 Difference Between Fast and Slow	0	1	0.095238
4 Availability	1	1	0.235294
5 Reliability	0.9	0.959	0.235294

IF0P0(FSD7) = 0.798

U/R FIGURES OF MERIT

REQUIREMENTS	UFX*i*P0(FSD7)	USF*i*P0(FSD7)	UW*i*P0
1 Acquisition Time	2	0.998	0.322581
2 Acquisition Cost	175	0.018	0.193548
3 Total Event Time	· 3.75	0.97	0.258065
4 Number of Electrical Circuits	1.5	0.119	0.096774
5 Number of Adults	6.5	0.182	0.129032

UF0P0(FSD7) = 0.611

5.5.2.8   *Figures of merit for Concepts 4 and 7 combined*
Concept 4 specifies a round-robin tournament with best-time scoring, and
Concept 7 specifies the use of electronic judging. A table for the prototype
method follows.

5.5.2.8.1   Prototype figures of merit for Concepts 4 and 7 combined (from the 1992 Derby, with 46 cars observed)

I/O FIGURES OF MERIT

REQUIREMENTS	IFXiP0(FSD7)	ISFiP0(FSD7)	IWiP0
1 Average Races per Car	6	0.935	0.147059
2 Number of Ties	0	1	0.088235
3 Happiness	0.771	0.907	0.294118
3.1 Percent Happy Scouts	92.6	0.74	0.238095
3.2 Number Irate Parents	0	1	0.142857
3.3 Number of Broken Cars	0	1	0.166667
3.4 Others Touching Scout's Car	1.47	0.744	0.095238
3.5 Number of Repeat Races	10	0	0.142857
3.6 Number of Lane Repeats	0	1	0.119048
3.7 Difference Between Fast and Slow	0	1	0.095238
4 Availability	1	1	0.235294
5 Reliability	0.8	0.929	0.235294

IF0P0(FSD7) = 0.946

U/R FIGURES OF MERIT

REQUIREMENTS	UFXiP0(FSD7)	USFiP0(FSD7)	UWiP0
1 Acquisition Time	10	0.97	0.322581
2 Acquisition Cost	72	0.98	0.193548
3 Total Event Time	3.5	1	0.258065
4 Number of Electrical Circuits	3	0	0.096774
5 Number of Adults	8	0.046	0.129032

UF0P0(FSD7) = 0.767

5.5.3   *Trade-off analysis*

The trade-off analysis compares the different design alternatives. After the figures of merit are collected and the scores computed, the Overall Perform-ance Figure of Merit and the Overall Utilization of Resources Figure of Merit are used to compute the trade-off scores for each category of figures of merit. Comparisons are made for the approximation, simulation, and prototype data. The symbology IF0P0(FSD1) indicates this is the Overall Input/Output Performance Figure of Merit for Problem 0 of Pinewood for the Functional System Design Concept 1.

5.5.3.1   *Approximation trade-off analysis*

The scores for the Input/Output Performance Requirement and the Utiliza-tion of Resources Requirement are summarized here with the Trade-Off Requirement.

5.5.3.1.1   *Trade-off scores*

Concept 1:   Single-elimination tournament

TW1P0 * IF0P0(FSD1) + TW2P0 * UF0P0(FSD1) = TF0P0(FSD1)
    0.9   *   0.656   +   0.1   *   0.786   =   0.668

Concept 2:   Double-elimination tournament

TW1P0 * IF0P0(FSD2) + TW2P0 * UF0P0(FSD2) = TF0P0(FSD2)
    0.9   *   0.669   +   0.1   *   0.656   =   0.669

Concept 3:   Round-robin tournament, mean-time scoring

TW1P0 * IF0P0(FSD3) + TW2P0 * UF0P0(FSD3) = TF0P0(FSD3)
    0.9   *   0.852   +   0.1   *   0.577   =   0.825

Concept 4:   Round-robin tournament, best-time scoring

TW1P0 * IF0P0(FSD4) + TW2P0 * UF0P0(FSD4) = TF0P0(FSD4)
    0.9   *   0.864   +   0.1   *   0.577   =   0.835

Concept 5:   Round-robin tournament, point-assignment scoring

TW1P0 * IF0P0(FSD5) + TW2P0 * UF0P0(FSD5) = TF0P0(FSD5)
    0.9   *   0.798   +   0.1   *   0.577   =   0.776

Concept 6:   Human judges

TW1P0 * IF0P0(FSD6) + TW2P0 * UF0P0(FSD6) = TF0P0(FSD6)
    0.9   *   0.601   +   0.1   *   0.798   =   0.621

Concept 7: Electronic judging

> TW1P0 * IF0P0(FSD7) + TW2P0 * UF0P0(FSD7) = TF0P0(FSD7)
>   0.9   *   0.725   +   0.1   *   0.685   =   0.721

5.5.3.1.2 Approximation alternatives.   The best alternative from the race formats (Concepts 1 through 5) and the best alternative from the judging (Concepts 6 and 7) will be combined into the overall optimal system design alternative. This is possible because these two sets of alternatives are independent of each other.

The best race format is Concept 4, the round-robin tournament using best-time scoring. The best judging alternative is Concept 7, the electronic system. These are based on guesses for the figures of merit and are used as the best concepts to begin focusing on.

Notice an anomaly in our scoring system. Concept 1, the single-elimination tournament got the same score as Concept 2, the double-elimination tournament. The Percent Happy Scouts was 50% for the first and 85% for the second, but the overall contribution of the scoring function was 0.0 and 0.119, respectively, times the weight of 0.238. We should have used this information to modify the scoring so that Percent Happy Scouts was emphasized more.

5.5.3.2 Simulation trade-off analysis

5.5.3.2.1 Trade-off scores

Concept 1: Single-elimination tournament

> TW1P0 * IF0P0(FSD1) + TW2P0 * UF0P0(FSD1) = TF0P0(FSD1)
>   0.9   *   0.637   +   0.1   *   0.786   =   0.650

Concept 2: Double-elimination tournament

> TW1P0 * IF0P0(FSD2) + TW2P0 * UF0P0(FSD2) = TF0P0(FSD2)
>   0.9   *   0.648   +   0.1   *   0.656   =   0.650

Concept 3: Round-robin tournament, mean-time scoring

> TW1P0 * IF0P0(FSD3) + TW2P0 * UF0P0(FSD3) = TF0P0(FSD3)
>   0.9   *   0.852   +   0.1   *   0.577   =   0.825

Concept 4: Round-robin tournament, best-time scoring

TW1P0 * IF0P0(FSD4) + TW2P0 * UF0P0(FSD4) = TF0P0(FSD4)
0.9   *   0.864   +   0.1 * 0.577   =   0.835

Concept 5: Round-robin tournament, point-assignment scoring

TW1P0 * IF0P0(FSD5) + TW2P0 * UF0P0(FSD5) = TF0P0(FSD5)
0.9 * 0.824   +   0.1 * 0.577   =   0.799

Concept 6: Human judges

TW1P0 * IF0P0(FSD6) + TW2P0 * UF0P0(FSD6) = TF0P0(FSD6)
0.9 * 0.528   +   0.1 * 0.767   =   0.552

Concept 7: Electronic judging

TW1P0 * IF0P0(FSD7) + TW2P0 * UF0P0(FSD7) = TF0P0(FSD7)
0.9 * 0.831   +   0.1 * 0.586   =   0.807

5.5.3.2.2 Simulation alternatives. The simulations were done on an IBM AT computer using Test Trajectory 1 for Concepts 1 through 5. Data for the races were based on 1991 actual races and were not varied (see Exhibit 5.3). The figures of merit Percent Happy Scouts and Number Irate Parents were estimated.

Simulations for Concepts 6 and 7 were done by randomizing data using Test Trajectory 1. The data for a round-robin format were used, and the data were varied using a normal data distribution (see Exhibit 5.4). An estimate for lane bias was created based on the actual data from 1991 (see Exhibit 5.5). Using these estimates, it was found that 12.9% of the races did not result in the fastest car winning. Most of this was the result of lane bias. Simulation estimated human judging errors were made in 5.2% of the races, with half of those from ties and half from calling the second place finisher the winner. The computer simulation, with a resolution to 0.0001 second, never made an error.

The best race format is Concept 4, the round-robin tournament using best-time scoring. The best judging method is Concept 7, the electronic system.

EXHIBIT 5.4

Statistical Race Data

Car	Average	Standard Deviation
A	2.5885	0.0194
B	2.6534	0.0158
C	2.6187	0.0300
D	2.6048	0.0193
E	2.5853	0.0090
F	2.6930	0.0278
G	2.6357	0.0241
H	2.8200	0.1773
I	3.0161	0.1531
J	2.6984	0.0363
K	2.7689	0.0241
L	2.8808	0.0637
M	3.0956	0.1300
N	2.6140	0.0138
O	2.6184	0.0241
P	2.8957	0.0536
Q	2.5818	0.0218
R	2.6583	0.0221
S	2.6130	0.0364
T	2.6050	0.0343
U	2.6593	0.0236
V	2.8664	0.0762
W	2.7743	0.0200

The data were assumed to be normally distributed. Each car was given a randomized finish time that was based on the average and standard deviation.

†EXHIBIT 5.5

Estimate of Lane Bias

Lane Number	Average	Standard Deviation	N	% of Lowest
1	2.7028	0.1436	46	1.0000
2	2.7357	0.1614	45	1.0122
3	2.7216	0.1616	46	1.0070
All	2.7203	0.1555	137	

The data were the results from races in the Webelos division. The data from another race (Bears division) showed similar results. Therefore, it was decided to include the lane bias as a percent increase over the true time of the car.

A confidence interval can be computed based on these measurements. The computations are shown below.

$$P\left(-1.96 \le \frac{2.7203 - \mu_T}{0.1436/\sqrt{45}} \le 1.96\right) = 0.95$$

The 95% confidence interval for the total of all lanes is then

$$C(2.6942 \le \mu_T \le 2.74226) = 0.95$$

which means that there is a 95% chance the mean is between these numbers. For lane 1,

$$C(2.6608 \le \mu_1 \le 2.7448) = 0.95$$

For lane 2,

$$C(2.6880 \le \mu_2 \le 2.7834) = 0.95$$

For lane 3,

$$C(2.6744 \le \mu_3 \le 2.7688) = 0.95$$

Examining the means of each lane, we see that no firm statement can be made regarding a lane bias, at least not with a 95% certainty. All the regions overlap, indicating they could all have the same time beyond some statistical variation. Indeed, by returning to the normal table we see that the data leave only a 70% confidence interval, which is not much confidence at all!

---

† Material in this exhibit is based on tools not presented in the text. It may be skipped without loss of continuity.

5.5.3.3 *Prototype trade-off analysis*

5.5.3.3.1 *Trade-off scores*

Concept 2:   Double-elimination tournament

$$TW1P0 * IF0P0(FSD2) + TW2P0 * UF0P0(FSD2) = TF0P0(FSD2)$$
$$0.9 \quad * \quad 0.628 \quad + \quad 0.1 \quad * \quad 0.842 \quad = \quad 0.649$$

Concept 4:   Round-robin tournament, best-time scoring

$$TW1P0 * IF0P0(FSD4) + TW2P0 * UF0P0(FSD4) = TF0P0(FSD4)$$
$$0.9 \quad * \quad 0.847 \quad + \quad 0.1 \quad * \quad 0.577 \quad = \quad 0.820$$

Concept 5:   Round-robin tournament, point-assignment scoring

$$TW1P0 * IF0P0(FSD5) + TW2P0 * UF0P0(FSD5) = TF0P0(FSD5)$$
$$0.9 \quad * \quad 0.804 \quad + \quad 0.1 \quad * \quad 0.615 \quad = \quad 0.785$$

Concept 6:   Human judges

$$TW1P0 * IF0P0(FSD6) + TW2P0 * UF0P0(FSD6) = TF0P0(FSD6)$$
$$0.9 \quad * \quad 0.541 \quad + \quad 0.1 \quad * \quad 0.761 \quad = \quad 0.563$$

Concept 7:   Electronic judging

$$TW1P0 * IF0P0(FSD7) + TW2P0 * UF0P0(FSD7) = TF0P0(FSD7)$$
$$0.9 \quad * \quad 0.798 \quad + \quad 0.1 \quad * \quad 0.611 \quad = \quad 0.779$$

5.5.3.3.2 *Prototype alternatives.*   Only prototypes from Concepts 2, 4, 5, 6, and 7 were built. Data for all of these concepts were available from prior years, thus historical data became our prototype. However, we have little confidence that the data presented were indeed collected as they were supposed to be. Dr. Bahill and Bill Karnavas have assured us that the quality of the data we received was acceptable. The only real surprise was the lack of reliability of the electronic scoring, which gave us a score of only 0.8. The reason for this was a brief system failure (a software error) during the race.

The best race format is Concept 4, the round robin using best-time scoring. The best judging method is Concept 7, the electronic system.

> In this design, the trade-off analysis produced the same conclusions
> for the approximation, simulation, and prototype data: The round-
> robin tournament with best-time scoring is the best race format and

electronic judging is better than human judging. It is unfortunate to have designs in which the three sets of data yield different conclusions because they will require expensive revisions.

5.5.4 Sensitivity analysis

The system is sensitive to the trade-off weightings. For example, changing the weights of the Trade-Off Requirement can easily sway the answer. The current trade-off puts heavy emphasis on the I/O performance of the system (0.90) and not on the utilization of resources (0.10). Changing the degree of emphasis can change the results, as summarized below using a 0.50/0.50 weighting and then a 0.30/0.70 weighting.

Weights are 0.50/0.50		Weights are 0.30/0.70	
Concept	Score	Concept	Score
2	0.735	2	0.778
4	0.712	4	0.658
5	0.709	5	0.672
6	0.651	6	0.695
7	0.704	7	0.667

In the 0.50/0.50 trade-off, the double-elimination format beats the round robin. This is because less time is spent in generating schedules and fewer adults are needed.

In the 0.30/0.70 trade-off, the double-elimination format is the best, as is the human judging alternative. Electronic judging loses because of its higher cost and greater use of time.

This indicates that if a scout pack is strapped for resources, the best approach is double-elimination with human judges. Otherwise, a round-robin format with electronic judging is the best system.

Bill Karnavas has done an extensive sensitivity analysis of this system (Karnavas et al., 1993). He found that only two parameters (out of 92) could change the recommended alternatives. The first was the trade-off weighting, as discussed above. The second was the slope of the scoring function for the figure of merit Percent Happy Scouts.

If this was increased from 0.1 to 0.3, Concept 3 (round robin, mean time) would be preferable to Concept 4 (round robin, best time).

This sensitivity study shows our design is insensitive to variations in almost all of the parameters. It is a robust design. We are pleased with this result.

5.5.5 Rationale for alternatives, models, and methods

An important part of systems engineering is encouraging an exploration of all possible alternatives. For the Pinewood Derby we briefly considered the following concepts:

1. Do not race; have a judge pick the winners based solely on appearance.

2. Race, but do not pick winners.

3. Have the audience vote on the winners by whatever criteria they choose.

4. Have every car race only once, with the fastest time winning.

5. Run handicap races. Measure times in initial races, then let the slower cars add weight.

6. Build a track with other than the traditional three lanes.

7. Run round-robin races, but arrange the schedules so that fast cars race fast cars and slow cars race slow cars.

8. Run a triple-elimination tournament.

We surveyed many techniques for deciding the winner of each race. The following five techniques received detailed analysis:

1. *Human observation.* This is the oldest and most common technique. Human judges are good at detecting the correct winner if the cars finish one or more inches apart (a 0.01-second difference). In closer races, humans often make mistakes or announce ties, which necessitates a subsequent rerun of the race.

2. *Photography.* A Polaroid camera could be mounted above the finish line to photograph the finish of each race. This would cost 75¢ per race and require one to two minutes for the photograph to develop. If the shutter were pressed at the wrong time, no cars would be in the field of view. This system was considered too slow, costly, and cumbersome.

3. *Bar code readers.* Paper bar codes could be glued to the bottom of each car, and bar code readers could be installed under the track at the finish line. This technique would not only tell which lane won, but also which car was in that lane. Merely stating that Lane 1 won could produce mistakes if, as often happens, Car A was supposed to be in Lane 1, but Car B was actually put there. The bar code readers we used cost $1000, and one would be needed for each lane. This alternative was considered too expensive.

4. *Optical sensors.* We used optical sensors mounted in the track at the start and finish lines to determine the winner of each race. The optical sensors were attached to electronic stop watches that were accurate to 0.01 second. We found that this was not more accurate than human judges. This system worked until the temperature dropped 30 °F, and the batteries lost their ability to deliver power. It has been said that such systems give false results in the presence of flash photography, although we did not experience this problem.

5. *Mechanical switches.* We installed mechanical switches in each lane at the start and finish lines. The disadvantages of such switches are that they bounce, and sometimes they fail to make good contact. However, we found ways to overcome these problems. The advantage of this mechanical switch timing system was that we could buy a complete system for $150. The mechanical switches were connected to an IBM-compatible personal computer. The system was accurate to 0.0001 second. We had no ties using this system. The computer was also used for scheduling and analyzing results.

For simplicity in the rest of this case study, the selection of the judging technique (Concepts 6 and 7) will not be considered a part of the system we are designing. We will only consider the consequences of 0.01-second and 0.0001-second resolutions.

5.6   Document 6:   System Functional Analysis

The System Functional Analysis Document decomposes the I/O
Requirements into a functional system design. Its intended audience
is systems engineers.

5.6.1   System functional analysis of Concept 1

5.6.1.1   Top level system functional analysis of Concept 1
System Concept 1 is a single-elimination tournament. The entire system has
been modeled based on the current design. The major components of this
model are shown in Figure 5.15. The major subfunctions are:

1. Inspect
2. Impound
3. Racing
4. Judging
5. Results

The system model shown in Figure 5.15 is the baseline that all other
concepts will alter. The modeling of the System Subfunction 5 (Results) is
altered for this concept.

5.6.1.2   Subfunction decomposition

5.6.1.2.1   Subfunction 1.   Subfunction 1 is Inspect. Cars enter the system
at this point. They are inspected for conformance to the rules of the Pinewood
Derby. If they pass, they proceed to the Impound area. If they fail, they leave
the system with a disqualified tag.

5.6.1.2.2   Subfunction 2.   Subfunction 2 is the Impound function. Cars
are placed in this holding area after they pass inspection and while they wait
for a race. They exit this area only on a request from the Racing component.

5.6.1.2.3   Subfunction 3.   Subfunction 3 is the Racing component. The
Racing component will perform the following functions:

1. If a new race then
      Get cars from the Impound area per the schedule and schedule
         index,
   else if a tie then
      Get cars from the Judging component based on the schedule and
         schedule index.
2. Set the cars at the starting blocks,

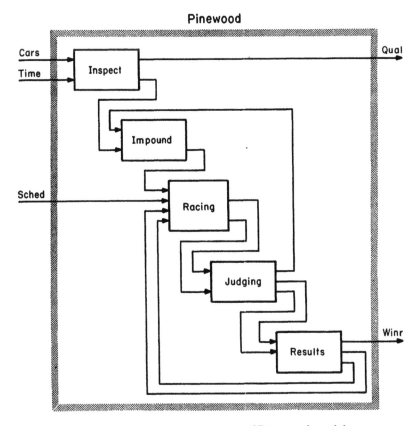

Figure 5.15 Major components of Pinewood model.

3. Start the race,
4. Send the cars back to the Judging component,
5. Output the new schedule index.

The schedule index is increased by one after each race. See Exhibit 5.6 for an example of a single-elimination tournament schedule. This schedule is input to the racing component and defines the scheduling of races. Cars are removed from the Impound component and placed in the appropriate lanes based on the schedule. The incrementing index tracks the races throughout the derby.

5.6.1.2.4 Subfunction 4. Subfunction 4 is the Judging component. The output of each race is sent to this component. The results of each race are decided as follows:

1. First place awarded to the car that crosses the finish line first.
2. Second place awarded to the car that crosses the finish line second.

EXHIBIT 5.6

Single-Elimination Tournament: 23 Cars (A Through W)

Race #	Lane 1	Lane 2	Lane 3	Comment
1	A	B	C	
2	D	E	F	
3	G	H	I	
4	J	K	L	
5	M	N	O	
6	P	Q	R	
7	S	T	U	
8	V	W		
9	F1	F2	F3	
10	F4	F5	F6	
11	F7	F8		
12	F9	F10	F11	1st is first place

Note: F1 means the first place finisher of the first race; F2 means the first place finisher of the second race; and so on.

3. Third place awarded to the car that crosses the finish line last.
4. A tie occurs if the first and second cars finish at the same time.
5. A nil occurs if the car does not cross the finish line.

The Judging component outputs the cars to the Impound area if there was no tie. If there was a tie, the cars are sent back to Racing component, and a judging flag is set to –1. If no judging is occurring, the flag is 0. If a valid race has occurred, then the judging flag is set to 1. The results of the race for each car are sent to the Results component.

5.6.1.2.5 Subfunction 5.   Subfunction 5 is the Results function. The results of each race are sent here from the Judging component. Race results are tallied per race. Results of every race are output to outside of the system (the spectators and scouts), showing the current place of each car.

In this particular example, the five subfunctions coincide with the five physical components of the system. This is not always the case; for example, a computer may handle hundreds of different functions on one processor.

5.6.1.3 *Complete subfunction model*

5.6.1.3.1 *Terminology*

```
Z1' = (SZ1', IZ1', OZ1', NZ1', RZ1')
```

where

  $Z1'$ = model of the Results component of the system,
 $SZ1'$ = states of the system,
 $IZ1'$ = inputs to the system,
 $OZ1'$ = outputs of the system,
 $NZ1'$ = next state function, and
 $RZ1'$ = readout function.

5.6.1.3.2 *States*

```
SZ1' = {Wait,FixSchedule#ijk,Tally#ijkp1p2p3}
```

This lists all the states where # is the index number; i j k represents the valid car names in Lanes 1, 2, and 3, respectively; and p1p2p3 are the places for Lanes 1, 2, and 3, respectively.

5.6.1.3.3 *Inputs*

```
IZ1' = I1Z1' × I2Z1'
  I1Z1' = {#,i,j,k,place1,place2,place3} where
            # = IJS[0,39] /*the schedule index*/
            i = ALPHA /*valid car label in
                         lane 1*/
            j = ALPHA /*valid car label in
                         lane 2*/
            k = ALPHA /*valid car label in
                         lane 3*/
            place1 = {1st,2nd,3rd,Nil}
                       /*finish place for car i*/
            place2 = {1st,2nd,3rd,Nil}
                       /*finish place for car j*/
            place3 = {1st,2nd,3rd,Nil}
                       /*finish place for car k*/
    I2Z1' = {(index,lane1,lane2,lane3)^num}
            /*this represents the schedule in*/
            /*the form of Exhibit 5.6. The*/
            /*variable num represents the*/
            /*length of the schedule*/
```

5.6.1.3.4 *Outputs*

```
OZ1' = O1Z1' × O2Z1' × O3Z1'
  O1Z1' = {index,i,j,k,place1,place2,place3}
```

```
                    /*This represents the Tally sheet*/
                    /*as shown in Exhibit 5.8*/
    02Z1' = IJS[1,INFINITY)   /*This is the*/
                    /*schedule index number*/
    03Z2' = {(index,lane1,lane2,lane3)^num}          '
                    /*This represents the new schedule*/
                    /*in the form of Exhibit 5.6. The*/
                    /*variable num represents the*/
                    /*length of the schedule.*/
```

5.6.1.3.5 Next state function

```
NZ1' = {((Wait,((0,x,x,x,x,x,x),any)),Wait),
         ((Wait,((#,i,j,k,place1,place2,place3),
             any)),FixSchedule#ijk),
         ((FixSchedule#ijk,((#,i,j,k,place1,
             place2,place3),any)),Tally#ijkp1p2p3),
         ((Tally#ijkp1p2p3,((#,i,j,k,place1,
             place2,place3),any)),Wait)}
```

where the next states, FixSchedule#ijk and Tally#ijkp1p2p3, correspond with the inputs #,i,j,k,place1,place2,place3 from Port 1. For example an input of (8,C,F,J,2nd,3rd,1st) would yield a next state of FixSchedule8CFJ or of Tally8CFJ2nd3rd1st.

5.6.1.3.6 Readout function

```
RZ1' = {(Wait,((0,nil^6),0,(0,0,0,0)^num)),
         (FixSchedule#ijk,((nil^6),nil,(#,i,j,k))),
         (Tally#ijkp1p2p3,((#,i,j,k,p1,p2,p3),
                         #+1,(nil^4)^num))}
```

5.6.2   System functional analysis of Concept 2

5.6.2.1   Top level system functional analysis of Concept 2
System Concept 2 is a double-elimination tournament. The entire system has been modeled on the current design. The major components of this model are the same as for Concept 1 and are shown in Figure 5.15. The modeling of the Results component from the baseline system is altered for this concept.

5.6.2.2   Subfunction decomposition
The system's subfunctions are decomposed the same as in Concept 1, except the schedule is different. Exhibit 5.7 is an example of a double-elimination tournament schedule. This schedule is provided to the Racing component and defines the scheduling of races.

EXHIBIT 5.7

Double-Elimination Tournament: 23 Cars (A through W)

Race #	Lane 1	Lane 2	Lane 3	Comment
1	A	B	C	
2.	D	E	F	
3	G	H	I	
4	J	K	L	
5	M	N	O	
6	P	Q	R	
7	S	T	U	
8	V	W		
9	F1	F2	F3	
10	F4	F5	F6	0 losses, 2nd race
11	F7	F8		0 losses, 2nd race
12	S1	S2	T7	1 loss, 2nd race
13	S3	S4	T6	1 loss, 2nd race
14	S5	T4	T3	1 loss, 2nd race
15	S6	S7	T5	1 loss, 2nd race
16	S8	T2	T1	1 loss, 2nd race
17	S9	T10	S11	1 loss, 3rd race
18	S10	T9	F12	1 loss, 3rd race
19	F13	F14		1 loss, 3rd race
20	F16	F15		1 loss, 3rd race
21	F9	F10	F11	0 losses, 3rd race, 1st is winner
22	S21	F17	F18	1 loss, 4th race
23	T21	F19	F20	1 loss, 4th race
24	F22	F23		1st is second, 2nd is third

Note: F1 is the first place finisher of the first race, F2 is the first place finisher of the second race, and so on. S1 is the second place finisher of the first race, and T1 is the third place finisher of the first race.

5.6.2.3   *Complete subfunction model*

5.6.2.3.1   *Terminology*

$$Z2' = (SZ2', IZ2', OZ2', NZ2', RZ2')$$

where

   $Z2'$ = model of the Racing component of Concept 2,
$SZ2'$ = states of the system,
$IZ2'$ = inputs to the system,

OZ2' = outputs of the system,
NZ2' = next state function, and
RZ2' = readout function.

This model is identical to that for Z1'.

5.6.3   *System functional analysis of Concept 3*

5.6.3.1   *Top level system functional analysis of Concept 3*

System Concept 3 is a round-robin format with mean-time scoring for the Results component. The entire system has been modeled on the current design. The major components of this model are the same as in Concept 1 and are shown in Figure 5.15. The modeling of the system Results component is altered for this concept.

5.6.3.2   *Subfunction decomposition*

The system is decomposed the same as in Concept 1, except the schedule is different. See Section 5.8 of this chapter for examples of round-robin tournament schedules. For this concept, the mean time of each race is calculated and stored in the Results subfunction. The result of each race is provided by the Judging component. The division winners are determined by the best mean score of each race.

5.6.3.3   *Complete subfunction model*

5.6.3.3.1   *Terminology*

Z3' = (SZ3', IZ3', OZ3', NZ3', RZ3')

where

Z3' = model of the Racing system,
SZ3' = states of the system,
IZ3' = inputs to the system,
OZ3' = outputs of the system,
NZ3' = next state function, and
RZ3' = readout function.

5.6.3.3.2   *States*

SZ3' = {Wait,FixSchedule#ijk,Tally#ijkp1p2p3}

5.6.3.3.3   *Inputs*

```
IZ3' = I1Z3' × I2Z3' × I3Z3'
  I1Z3' = {#,i,j,k,place1,place2,place3} where
          # = IJS[0,39] /*the schedule index*/
          i = ALPHA /*valid car label in lane 1*/
          j = ALPHA /*valid car label in lane 2*/
          k = ALPHA /*valid car label in lane 3*/
```

EXHIBIT 5.8

Part of a Tally Sheet

Webelos

			Round Number								
	Pack 212 Pinewood Derby Tally Sheet										
Car Label	Scout's Name	Den	1	2	3	4	5	6	Result	Den Winners	Division Winners
A											
B											
C											
D											
E											
F											
G											
•											
•											
•											
DD											
EE											
FF											
GG											
HH											
II											
JJ											

```
          place1 = IJS[0,INFINITY)
                   /*finish place for car i*/
          place2 = IJS[0,INFINITY)
                   /*finish place for car j*/
          place3 = IJS[0,INFINITY)
                   /*finish place for car k*/
    I2Z3' = {(index,lane1,lane2,lane3)^num}
            /*this represents the schedule in*/
            /*the form of Exhibit 5.6. The*/
            /*variable num represents the*/
            /*length of the schedule*/
```

5.6.3.3.4 Outputs

```
0Z3' = 01Z3' × 02Z3' × 03Z3'
   01Z3' = {index,i,j,k,place1,place2,place3}
                /*This represents the Tally sheet*/
                /*as shown in Exhibit 5.8*/
   02Z3' = IJS[1,INFINITY)  /*This is the*/
                /*schedule index number*/
   03Z3' = {(index,lane1,lane2,lane3)^num}
                /*This represents the new schedule*/
                /*in the form of those shown in*/
                /*Section 5.8. The variable num*/
                /*represents the length of the*/
                /*schedule.*/
```

5.6.3.3.5 Next state function

```
NZ3' = {((Wait,((0,x,x,x,x,x,x),any)),Wait),
         ((Wait,((#,i,j,k,place1,place2,place3),
              any)),FixSchedule#ijk),
         ((FixSchedule#ijk,((#,i,j,k,place1,
              place2,place3),any)),Tally#ijkp1-p2-p3),
         ((Tally#ijkp1p2p3,((#,i,j,k,place1,
              place2,place3),any)),Wait)}
```

where the next states FixSchedule#ijk and Tally#ijkp1p2p3 correspond with the inputs #,i,j,k,place1,place2,place3 from Port 1. For example, an input of (8,C,F,J,40,43,35) means the eighth race per the schedule using Cars C, F, and J resulted in times of 40, 43, and 35, respectively. This would yield a next state of FixSchedule8CFJ to update the schedule and then Tally8CFJ40-43-35 to update the tally sheets.

5.6.3.3.6 Readout function

```
RZ3' = {(Wait,((0,nil^6),0,(0,0,0,0)^num)),
         (FixSchedule#ijk,((nil^6),nil,(#,i,j,k))),
         (Tally#ijkp1-p2-p3,((#,i,j,k,p1,p2,p3),
                            #+1,(nil^4)^num))}
```

5.6.4 System functional analysis of Concept 4

5.6.4.1 Top level system functional analysis of Concept 4

System Concept 4 is a round-robin format, the winner being determined by the fastest race time. The entire system has been modeled on the current design. The major components of this model are identical to Concept 3 except for the Results section.

5.6.4.2 Subfunction decomposition

The functional decomposition is the same as Concept 3 except for the Results subfunction. For this concept, the best time in each race is calculated and stored in the Results component. The result of each race is as provided by the Judging component. The division winners are those having the best time in all the races.

5.6.4.3 Complete subfunction model

5.6.4.3.1 Terminology

$$Z4' = (SZ4', IZ4', OZ4', NZ4', RZ4')$$

where

$Z4'$ = model of the system,
$SZ4'$ = states of the system,
$IZ4'$ = inputs to the system,
$OZ4'$ = outputs of the system,
$NZ4'$ = next state function, and
$RZ4'$ = readout function.

$Z4'$ is identical to $Z3'$ except for the scoring method used.

5.6.5 System functional analysis of Concept 5

5.6.5.1 Top level system functional analysis of Concept 5

System Concept 5 is a round-robin format with point-assignment scoring. The entire system has been modeled on the current design. The major components of this model are identical to Concept 3 except that system components Racing and Results are altered for this concept.

5.6.5.2 Subfunction decomposition

5.6.5.2.1 *Subfunction 1.* Subfunction 1 is the Inspect function. This is the same as for Concept 3.

5.6.5.2.2 *Subfunction 2.* Subfunction 2 is the Impound function. This is same as for Concept 3.

5.6.5.2.3 *Subfunction 3.* Subfunction 3 is the Racing component. The Racing component will perform the following functions:

1. If a new race then
    Get cars from the Impound area per the schedule and schedule index,
  else if a tie then
    Get cars from the Judging component for the schedule and schedule index.

2. Set the cars at the starting blocks,
3. Start the race,
4. Send the cars back to the Judging component,
5. Output the new schedule index.

The schedule index is increased by one each time. See Section 5.8 for examples of round-robin tournament schedules. One of these schedules is input to the Racing component and defines the scheduling of races. Cars are removed from the Impound component and placed in the appropriate lanes based on the schedule. The incrementing index tracks the races throughout the derby.

5.6.5.2.4 Subfunction 4.   Subfunction 4 is the Judging component. This component is the same as for Concept 3.

5.6.5.2.5 Subfunction 5.   Subfunction 5 is the Results function. The results of each race are sent here from the Judging component. Race results are tallied per the race, the pack, and the division. Results are output external to the system (the spectators and scouts), clearly showing the current place of each car.

For this concept, each race score is calculated and stored based on 3 points for first, 2 points for second, 1 point for third, and 0 points for a no-show or a race not completed. The result of each race is as provided by the Judging component. The division winners are determined by the best average score of each race.

5.6.5.3   Complete subfunction model

5.6.5.3.1   Terminology

$$Z5' = (SZ5', IZ5', OZ5', NZ5', RZ5')$$

where

> $Z5'$ = model of the Racing component,
> $SZ5'$ = states of the system,
> $IZ5'$ = inputs to the system,
> $OZ5'$ = outputs of the system,
> $NZ5'$ = next state function, and
> $RZ5'$ = readout function.

System $Z5'$ is identical to system $Z3'$ except for the scoring method.

5.6.6   *System functional analysis of Concept 6*

5.6.6.1   *Top level system functional analysis of Concept 6*
System Concept 6 uses a human judge to decide winners of races. The entire system has been modeled on the current design. The major components of this model are the same as in Concept 1, as shown in Figure 5.15. The modeling of the system component Judging is altered for this concept.

5.6.6.2   *Subfunction decomposition*

The system decomposition is the same for this model as that for Concept 1, except for the Judging component. The judges will decide which car has won only when the difference in their finish times is greater than 0.01 s. Otherwise, a tie will be declared.

5.6.6.3   *Complete subfunction model*

5.6.6.3.1   *Terminology*

```
Z6' = (SZ6', IZ6', OZ6', NZ6', RZ6')
```

where

    $Z6'$ = model of the Judging system ,
    $SZ6'$ = states of the system,
    $IZ6'$ = inputs to the system,
    $OZ6'$ = outputs of the system,
    $NZ6'$ = next state function, and
    $RZ6'$ = readout function.

5.6.6.3.2   *States*

```
SZ6' = {Start,Lane1First,Lane2First,Lane3First,
        Lane123ijk,Lane132ijk,Lane213ijk,
        Lane231ijk,Lane312ijk,Lane321ijk,Tie}
```

where i j k represents the valid names of cars in lanes 1, 2, and 3, respectively.

5.6.6.3.3   *Inputs*

```
IZ6' = {(car,t)^3} /*where car is any valid*/
                   /*name of a car in the*/
                   /*derby or is Nil, and t is*/
                   /*the time the car reached*/
                   /*the finish line*/
```

5.6.6.3.4   *Outputs*

```
OZ6' = O1Z6' × O2Z6'
  O1Z6' = {cars^3} /*where cars represents any*/
                   /*valid name of a car in*/
                   /*the derby, or is Nil*/
  O2Z6' = {-1,0,1} /*where -1 represents a*/
                   /*tie, 0 is no race, and 1*/
                   /*is a valid race*/
    O3Z6' = {(car,place)^3} /*where car is a*/
                   /*valid car entry and place*/
                   /*is First, Second, Third,*/
                   /*Tie, or Nil*/
```

```
O4Z6' = {cars^3} /*where cars represents any*/
                 /*valid name of a car in*/
                 /*the derby, or is Nil*/
```

5.6.6.3.5 *Next state function*

```
NZ6' = {((Start,f1),NextState1),
        ((Lane1First,f2),NextState2),
        ((Lane2First,f3),NextState3),
        ((Lane3First,f4),NextState4),
        ((Lane123ijk,any),Start),
        ((Lane132ijk,any),Start),
        ((Lane213ijk,any),Start),
        ((Lane231ijk,any),Start),
        ((Lane312ijk,any),Start),
        ((Lane321ijk,any),Start),
        ((Tie,any),Start)}
```

where

```
/*f1 determines who is first*/

f1 = (let ((p11,p12),(p21,p22),(p31,p32))=IZ6';
      if (p12 > p22+resolve and p12 >
          p32+resolve) then
          NextState1 = Lane1First
      else if (p22 > p12+resolve and p22 >
          p32+resolve) then
          NextState1 = Lane2First
      else if (p32 > p12+resolve and p32 >
          p22+resolve) then
          NextState1 = Lane3First
      else if (p11=Nil and p21=Nil and
               P31=Nil) then
          NextState1 = Start
      else
          NextState1 = Tie;)

/*f2 determines who is second and third if 1*/
/*is first*/

f2 = (let ((p11,p12),(p21,p22),(p31,p32))=IZ6';
      if (p22 > p32+resolve) then
          NextState2 = Lane123ijk
      else
          NextState2 = Lane132ijk;)
```

```
/*f3 determines who is second and third if 2*/
/*is first*/

f3 = (let ((p11,p12),(p21,p22),(p31,p32))=IZ6';
      if (p12 > p32 +resolve) then
         NextState3 = Lane213ijk
      else
         NextState3 = Lane231ijk;)

/*f4 determines who is second and third if 3*/
/*is first*/

f4 = (let ((p11,p12),(p21,p22),(p31,p32))=IZ6';
      if (p12 > p22 +resolve) then
         NextState4 = Lane312ijk
      else
         NextState4 = Lane321ijk;)
```

where resolve = 0.01 for human judges, and i j k corresponds to p11, p21, and p31, respectively.

5.6.6.3.6 Readout function

```
RZ6' = {(Start,((Nil)^3,0,(Nil,Nil)^3)),(Nil)^3),
        (Lane1First,((Nil)^3,0,(Nil,Nil)^3),
                     (Nil)^3),
        (Lane2First,((Nil)^3,0,(Nil,Nil)^3),
                     (Nil)^3),
        (Lane3First,((Nil)^3,0,(Nil,Nil)^3),
                     (Nil)^3),
        (Lane123ijk,((Nil)^3,1,((p11,First),
         (p21,Second),(p31,Third)),(i,j,k)),
        (Lane132ijk,((Nil)^3,1,((p11,First),
         (p21,Third),(p31,Second)),(i,j,k)),
        (Lane213ijk,((Nil)^3,1,((p11,Second),
         (p21,First),(p31,Third)),(i,j,k)),
        (Lane231ijk,((Nil)^3,1,((p11,Third),
         (p21,First),(p31,Second)),(i,j,k)),
        (Lane312ijk,((Nil)^3,1,((p11,Second),
         (p21,Third),(p31,First)),(i,j,k)),
        (Lane321ijk,((Nil)^3,1,((p11,Third),
         (p21,Second),(p31,First)),(i,j,k)),
        (Tieijk,(ijk,-1,(Nil,Nil)^3),(Nil)^3)}
```

where we let ((p11,p12),(p21,p22),(p31,p32)) = IZ6'.

5.6.7   *System functional analysis of Concept 7*

5.6.7.1   *Top level system functional analysis of Concept 7*
System Concept 7 uses an electronic system to judge the winners of races. The entire system has been modeled on the current design. The major components of this model are the same as for Concept 5. The modeling of the Judging system component is altered for this concept.

5.6.7.2   *Subfunction decomposition*
The subfunction decomposition is identical to that for Concept 6 except that the Judging component is altered. The resolution (Resolve in the model) is 0.0001 s. If the difference in time between cars passing the finish line are less than the resolution, there is a tie; otherwise, a winner is declared.

5.7   *Document 7:   System Physical Synthesis*

> *The System Physical Synthesis Document develops and explains the relationships between the models of the previous documents and the physical components that will comprise the final system. It is created in conjunction with Document 6.*

5.7.1   *Physical synthesis of Concept 1*

5.7.1.1   *Top level system design of Concept 1*
System Concept 1 is for a single-elimination tournament. Concepts 1 through 5 differ only in the Results component of the functional design. The original system will continue unaltered with the exception of this change.

The physical decomposition will be as follows:

1. A judging system (determined by Concepts 6 and 7).
2. A paper schedule of races will be provided.
3. A paper tally sheet will be provided.

5.7.1.2   *Subunit physical synthesis*

5.7.1.2.1 Subunit 1.   At the end of each race, the first, second, and third place winners will be determined. The names of the cars in the race and the results are combined for one input. The other inputs are the schedule and race index. The place the participants finish in will be recorded in the Results column of the race schedule and in the tally sheet, as shown in Exhibit 5.8. The winner of each race will be designated as F#, where # is the index number.

The second place finisher will be S#, and the third place finisher, T#. The schedule is updated to indicate these results. The tally sheet will be updated with the results of this race, and the results will also be made available to the participants.

5.7.1.2.2 *Subunit 2.* A paper schedule of races will be provided. A sample of this schedule for a single-elimination tournament is given in Exhibit 5.6.

5.7.1.2.3 *Subunit 3.* A Tally sheet will be used for this race as per Exhibit 5.8.

5.7.2 Physical synthesis of Concept 2

5.7.2.1 Top level system design of Concept 2
System Concept 2 is for a double-elimination tournament. This affected only the Results component of the functional design. The original system will continue unaltered with the exception of this change.

The physical decomposition will be the same as for Concept 1, except the schedule is different. See Exhibit 5.7 for an example.

5.7.3 Physical synthesis of Concept 3

5.7.3.1 Top level system design of Concept 3
System Concept 3 is for a round-robin tournament with mean-time scoring. This affected only the Results component of the functional design. The original system will continue unaltered with the exception of this change.

5.7.3.2 Subunit physical synthesis

5.7.3.2.1 *Subunit 1.* At the end of each race, the first, second, and third place winners will be determined. The names of the cars in the race and the results are combined for one input. The other inputs are the schedule and race index. The actual finish times will be recorded in the Results column of the race schedule and in the tally sheet, as shown in Exhibit 5.8. The winner of each race will be designated as F#, where # is the index number. The second place finisher will be S#, and the third place finisher, T#. The schedule is updated to indicate these results. The tally sheet will be updated with the results of this race, and the results will also be made available to the participants. The winner of all races will be determined at the end of all races. At that time, an average of all race times will be calculated. In this unit, only the times are recorded.

5.7.3.2.2 *Subunit 2.* A paper schedule of races will be provided. Sample schedules for a round-robin tournament are given in Section 5.8.

5.7.3.2.3 *Subunit 3.* A Tally sheet will be used for this race as per Exhibit 5.8.

5.7.4 *Physical synthesis of Concept 4*

5.7.4.1 *Top level system design of Concept 4*
System Concept 4 is for a round-robin tournament with best-time scoring. This system is functionally identical to Concept 3 except for the Results component of the functional design

5.7.4.2 *Subunit physical synthesis*

5.7.4.2.1 *Subunit 1.*   At the end of each race, the first, second, and third place winners will be determined. The names of the cars in the race and the results are combined for one input. The other inputs are the schedule and race index. The actual finish times will be recorded in the Results column of the race schedule and in the tally sheet, as shown in Exhibit 5.8. The winner of each race will be designated as F#, where # is the index number. The second place finisher will be S#, and the third place finisher, T#. The schedule will be updated to indicate these results. The tally sheet will be updated with the results of this race and the results will also be made available to the participants. The winner of all races will be determined at the end of all races. At that time, the best time of all the races for each participant will be calculated. In this unit, only the times will be recorded.

5.7.4.2.2 *Subunit 2.*   This subunit is identical to Subunit 2 of Concept 3.

5.7.4.2.3 *Subunit 3.*   This subunit is identical to Subunit 3 of Concept 3.

5.7.5 *Physical synthesis of Concept 5*

5.7.5.1 *Top level system design of Concept 5*
System Concept 5 is for a round-robin tournament with point-assignment scoring. This system is functionally identical to Concept 3 except for the Results component of the functional design.

5.7.5.2 *Subunit physical synthesis*

5.7.5.2.1 *Subunit 1.*   At the end of each race, the first, second, and third place winners will be determined. The names of the cars in the race and the result are combined for one input. The other inputs are the schedule and race index. The actual finish times will be recorded in the results column of the race schedule and in the tally sheet, as shown in Exhibit 5.8. The winner of each race will be designated as F#, where # is the index number. The second place finisher will be S#, and the third place finisher, T#. The schedule will be updated to indicate these results. The tally sheet will be updated with the results of this race and the results will also be made available to the participants. The winner of all races will be determined at the end of all races. The winner of a race receives 1 point, second place receives 2 points, and third place receives 3 points. At the completion of all races the winners are the ones with the lowest overall sum of points.

5.7.5.2.2 *Subunit 2.* This subunit is identical to Subunit 2 of Concept 3.

5.7.5.2.3 *Subunit 3.* This subunit is identical to Subunit 3 of Concept 3.

5.7.6 Physical synthesis of Concept 6

5.7.6.1 Top level system design of Concept 6

System Concept 6 specifies a human judge to determine winners. This affects only the Judging component of the functional design. The original system will continue unaltered with the exception of this change.

The physical decomposition will be as follows:

1. Two persons will be used: Judge 1 and Judge 2.
2. A paper schedule of races will be provided.

5.7.6.2 Subunit physical synthesis

5.7.6.2.1 *Subunit 1.* The primary jobs of Judge 1 and Judge 2 are to determine the winners. An additional job is to control the crowd. The Finish Line Judge (Judge 1) will watch the finish line and (1) ensure that the cars are in the proper lanes and (2) reset the finish line switches and tell the starter when they are ready for the next race. The Finish Line Facilitator (Judge 2) will keep the scouts away from the finish line. After the Finish Line Judge has observed the race and reset the switches, the Finish Line Facilitator will pick up the cars and hand them to the scouts or, if the scout owner is not there, put them on a pillow.

At the end of each race, Judge 1 calls out the first, second, and third place lane numbers. In other words, if the fastest car was in Lane 2, the second place car was in Lane 1, and the slowest car was in Lane 3, the judge would call out, "Two, one, three."

5.7.6.2.2 *Subunit 2.* Paper schedules of races will be used as for Concepts 1, 2, and 3.

5.7.7 Physical synthesis of Concept 7

5.7.7.1 Top level system design of Concept 7

System Concept 7 specifies an electronic system to determine winners. This affects only the Judging component of the functional design. The original system will continue unaltered with the exception of this change.

The physical decomposition will be as follows:

1. Two persons will be used: a Finish Line Judge and a Computer "Guru."
2. A paper schedule of races will be provided.
3. Sensors are connected to the end of the racetrack and interfaced to a personal computer with appropriate software.

5.7.7.2 Subunit physical synthesis

5.7.7.2.1 Subunit 1.   The race will be computerized. The jobs of the Finish Line Judge are to control the crowds, reset the finish line switches, verify that the computer is working correctly, and be prepared to step in and run the race manually in the case of power failure. The Finish Line Judge will watch the finish line and (1) ensure that the cars are in the proper lanes, (2) reset the finish line sensors and tell the starter when they are ready for the next race, and (3) keep the scouts away from the finish line. The Finish Line Judge will then pick up the cars and hand them to the scouts or, if the scout owner is not there, put them on a pillow. The Computer Guru will be available to troubleshoot in case of computer malfunction.

5.7.7.2.2 Subunit 2.   A paper schedule of races, as shown in the exhibits for Concepts 1, 2, and 3, will be used.

5.7.7.2.3 Subunit 3.   Sensors that detect the passage of the cars will be installed at the end of the racetrack. These will be interfaced to a computer with software that can determine race time. The judge must reset these sensors after each race. The sensors are capable of determining the race times to a resolution of 0.0001 second.

5.8 Round-robin schedules for a Pinewood Derby

Cub scouts enjoy racing their cars against their friends, and they want these races to be fair. The following schedules were constructed with a view toward making the scouts happy. In these schedules, each car races six times, twice in each of the three lanes, and whenever possible, no scout races any other scout twice.

If your derby has fewer cars than are listed in the schedules, just ignore the missing cars. For example, in an eight car round robin, ignore the car labeled "I." For a seven car round robin, ignore Cars H and I; Car G still runs the first race, even though it is unopposed. It is important that each car run the same number of races to avoid unfair warm-up or wear-out effects and to allow each car six tries to demonstrate its fastest time.

For the following schedules, the first round was arranged by inspection time to allow the scouts some control over their destinies. Later rounds were derived using a random number generator to ensure fairness. The important conditions are that each car races six times and each car runs in each lane twice.

In nine and twelve car round robins with six races for each car, each car races every other car at least once. For round robins of 15 or more cars, each car races in each lane twice and no two cars race each other twice, but obviously every car cannot race every other car.

The schedules in this section can also be used as tally sheets to record the results. They were computed with great effort by William J. Karnavas.

9 Car Round-Robin Schedule

	Lane 1		Lane 2		Lane 3	
	Car	Place	Car	Place	Car	Place
Round 1						
Race 1	A		B		C	
Race 2	D		E		F	
Race 3	G		H		I	
Round 2						
Race 1	A		D		G	
Race 2	B		E		H	
Race 3	C		F		I	
Round 3						
Race 1	B		F		G	
Race 2	D		C		H	
Race 3	E		I		A	
Round 4						
Race 1	C		G		E	
Race 2	F		H		A	
Race 3	I		D		B	
Round 5						
Race 1	E		B		C	
Race 2	G		A		F	
Race 3	H		I		D	
Round 6						
Race 1	F		C		D	
Race 2	H		A		E	
Race 3	I		G		B	

12 Car Round-Robin Schedule

	Lane 1		Lane 2		Lane 3	
	Car	Place	Car	Place	Car	Place

Round 1

Race 1	A		B		C	
Race 2	D		E		F	
Race 3	G		H		I	
Race 4	J		K		L	

Round 2

Race 1	B		D		H	
Race 2	A		E		J	
Race 3	G		C		K	
Race 4	F		I		L	

Round 3

Race 1	C		H		E	
Race 2	D		I		J	
Race 3	B		F		K	
Race 4	L		A		G	

Round 4

Race 1	H		L		I	
Race 2	E		G		B	
Race 3	C		F		J	
Race 4	K		A		D	

Round 5

Race 1	J		B		G	
Race 2	H		F		A	
Race 3	I		K		E	
Race 4	L		D		C	

Round 6

Race 1	K		J		H	
Race 2	E		L		B	
Race 3	I		C		A	
Race 4	F		G		D	

Note: This schedule is not perfect because Car B does not race Car I, but it is the best schedule we could find.

15 Car Round-Robin Schedule

	Lane 1		Lane 2		Lane 3	
	Car	Place	Car	Place	Car	Place

Round 1

Race 1	A		B		C	
Race 2	D		E		F	
Race 3	G		H		I	
Race 4	J		K		L	
Race 5	M		N		O	

Round 2

Race 1	C		F		G	
Race 2	D		B		H	
Race 3	A		J		M	
Race 4	E		K		N	
Race 5	I		L		O	

Round 3

Race 1	L		N		C	
Race 2	E		G		J	
Race 3	H		F		K	
Race 4	I		M		B	
Race 5	O		A		D	

Round 4

Race 1	M		C		D	
Race 2	B		E		L	
Race 3	G		A		K	
Race 4	J		O		H	
Race 5	F		I		N	

Round 5

Race 1	N		D		G	
Race 2	F		L		A	
Race 3	C		I		J	
Race 4	H		M		E	
Race 5	K		O		B	

Round 6

Race 1	N		H		A	
Race 2	B		J		F	
Race 3	K		D		I	
Race 4	L		G		M	
Race 5	O		C		E	

18 Car Round-Robin Schedule

	Lane 1		Lane 2		Lane 3	
	Car	Place	Car	Place	Car	Place

Round 1

Race 1	A		B		C	
Race 2	D		E		F	
Race 3	G		H		I	
Race 4	J		K		L	
Race 5	M		N		O	
Race 6	P		Q		R	

Round 2

Race 1	E		N		I	
Race 2	R		O		A	
Race 3	H		B		K	
Race 4	C		J		P	
Race 5	Q		F		G	
Race 6	L		D		M	

Round 3

Race 1	B		L		N	
Race 2	O		P		D	
Race 3	E		G		C	
Race 4	M		F		R	
Race 5	H		J		A	
Race 6	I		K		Q	

Round 4

Race 1	P		E		B	
Race 2	I		A		L	
Race 3	N		Q		D	
Race 4	F		O		H	
Race 5	R		C		K	
Race 6	G		M		J	

18 Car Round-Robin Schedule *continued*

	Lane 1		Lane 2		Lane 3	
	Car	Place	Car	Place	Car	Place
Round 5						
Race 1	A		G		N	
Race 2	J		I		F	
Race 3	D		R		B	
Race 4	L		P		H	
Race 5	O		C		Q	
Race 6	K		M		E	
Round 6						
Race 1	Q		L		E	
Race 2	B		I		O	
Race 3	C		H		M	
Race 4	N		R		J	
Race 5	F		A		P	
Race 6	K		D		G	

21 Car Round-Robin Schedule

	Lane 1		Lane 2		Lane 3	
	Car	Place	Car	Place	Car	Place

Round 1

Race 1	A		B		C	
Race 2	D		E		F	
Race 3	G		H		I	
Race 4	J		K		L	
Race 5	M		N		O	
Race 6	P		Q		R	
Race 7	S		T		U	

Round 2

Race 1	I		Q		J	
Race 2	F		O		H	
Race 3	M		P		S	
Race 4	R		T		C	
Race 5	U		A		D	
Race 6	K		E		B	
Race 7	G		L		N	

Round 3

Race 1	U		F		I	
Race 2	C		D		G	
Race 3	T		J		E	
Race 4	A		H		R	
Race 5	K		S		N	
Race 6	O		L		P	
Race 7	B		M		Q	

Round 4

Race 1	L		U		B	
Race 2	I		C		E	
Race 3	N		P		T	
Race 4	D		J		O	
Race 5	S		A		Q	
Race 6	R		F		G	
Race 7	H		K		M	

21 Car Round-Robin Schedule *continued*

	Lane 1		Lane 2		Lane 3	
	Car	Place	Car	Place	Car	Place
			Round 5			
Race 1	T		M		L	
Race 2	B		R		D	
Race·3	Q		U		K	
Race 4	N		C		F	
Race 5	E		G		P	
Race 6	H		S		J	
Race 7	O		I		A	
			Round 6			
Race 1	C		O		U	
Race 2	J		N		A	
Race 3	P		I		K	
Race 4	L		D		H	
Race 5	E		R		M	
Race 6	F		B		S	
Race 7	Q		G		T	

24 Car Round-Robin Schedule

	Lane 1		Lane 2		Lane 3	
	Car	Place	Car	Place	Car	Place

Round 1

Race 1	A		B		C	
Race 2	D		E		F	
Race 3	G		H		I	
Race 4	J		K		L	
Race 5	M	.	N		O	
Race 6	P		Q		R	
Race 7	S		T		U	
Race 8	V		W		X	

Round 2

Race 1	L		P		F	
Race 2	B		M		Q	
Race 3	H		W		A	
Race 4	T		C		V	
Race 5	S		D		X	
Race 6	I		E		N	
Race 7	G		J		U	
Race 8	R		K		O	

Round 3

Race 1	X		L		G	
Race 2	W		N		Q	
Race 3	M		J		H	
Race 4	A		F		I	
Race 5	R		S		B	
Race 6	C		O		P	
Race 7	K		T		D	
Race 8	E		U		V	

Round 4

Race 1	N		S		P	
Race 2	F		B		W	
Race 3	Q		O		G	
Race 4	J		R		T	
Race 5	U		X		H	
Race 6	V		A		D	
Race 7	I		L		C	
Race 8	E		M		K	

24 Car Round-Robin Schedule *continued*

	Lane 1		Lane 2		Lane 3	
	Car	Place	Car	Place	Car	Place

Round 5

	Car	Place	Car	Place	Car	Place
Race 1	W		C		E	
Race 2	F		Q		J	
Race 3	U		P		B	
Race 4	X		I		K	
Race 5	D		G		M	
Race 6	T		A		L	
Race 7	H		R		N	
Race 8	O		V		S	

Round 6

	Car	Place	Car	Place	Car	Place
Race 1	K		U		W	
Race 2	L		V		M	
Race 3	N		X		A	
Race 4	B		D		J	
Race 5	O		F		T	
Race 6	C		G		R	
Race 7	P		H		E	
Race 8	Q		I		S	

27 Car Round-Robin Schedule

	Lane 1		Lane 2		Lane 3	
	Car	Place	Car	Place	Car	Place
Round 1						
Race 1	A		B		C	
Race 2	D		E		F	
Race 3	G		H		I	
Race 4	J		K		L	
Race 5	M		N		O	
Race 6	P		Q		R	
Race 7	S		T		U	
Race 8	V		W		X	
Race 9	Y		Z		AA	
Round 2						
Race 1	L		R		S	
Race 2	M		A		Y	
Race 3	U		I		W	
Race 4	J		Z		X	
Race 5	B		T		V	
Race 6	H		AA		C	
Race 7	Q		D		G	
Race 8	E		O		K	
Race 9	F		N		P	
Round 3						
Race 1	N		R		I	
Race 2	Z		A		W	
Race 3	C		U		V	
Race 4	B		X		K	
Race 5	F		Q		AA	
Race 6	L		Y		T	
Race 7	D		P		H	
Race 8	O		G		J	
Race 9	S		E		M	
Round 4						
Race 1	G		M		F	
Race 2	R		O		T	
Race 3	P		V		Y	
Race 4	H		J		Q	
Race 5	A		S		D	
Race 6	I		X		L	
Race 7	K		C		Z	
Race 8	W		AA		B	
Race 9	N		U		E	

	Lane 1		Lane 2		Lane 3	
	Car	Place	Car	Place	Car	Place
Round 5						
Race 1	X		S		G	
Race 2	T		P		Z	
Race 3	I		Y		J	
Race 4	Q		W		N	
Race 5	AA		D		M	
Race 6	C		F		O	
Race 7	E		B		H	
Race 8	R		K		U	
Race 9	V		L		A	
Round 6						
Race 1	U		J		A	
Race 2	K		M		P	
Race 3	O		V		Q	
Race 4	T		C		D	
Race 5	W		L		N	
Race 6	X		H		R	
Race 7	Y		F		B	
Race 8	Z		G		E	
Race 9	AA		I		S	

27 Car Round-Robin Schedule *continued*

30 Car Round-Robin Schedule

	Lane 1		Lane 2		Lane 3	
	Car	Place	Car	Place	Car	Place
Round 1						
Race 1	A		B		C	
Race 2	D		E		F	
Race 3	G		H		I	
Race 4	J		K		L	
Race 5	M		N		O	
Race 6	P		Q		R	
Race 7	S		T		U	
Race 8	V		W		X	
Race 9	Y		Z		AA	
Race 10	BB		CC		DD	
Round 2						
Race 1	W		CC		A	
Race 2	Y		C		X	
Race 3	Z		B		BB	
Race 4	K		F		Q	
Race 5	DD		L		AA	
Race 6	U		D		R	
Race 7	E		M		H	
Race 8	G		S		V	
Race 9	P		N		J	
Race 10	I		O		T	
Round 3						
Race 1	U		M		B	
Race 2	CC		S		H	
Race 3	N		V		C	
Race 4	BB		O		J	
Race 5	Q		W		T	
Race 6	I		D		DD	
Race 7	X		R		K	
Race 8	L		P		Y	
Race 9	Z		A		E	
Race 10	F		AA		G	

30 Car Round-Robin Schedule *continued*

	Lane 1		Lane 2		Lane 3	
	Car	Place	Car	Place	Car	Place
Round 4						
Race 1	Q		Y		BB	
Race 2	D		V		P	
Race 3	W		G		B	
Race 4	L		H		U	
Race 5	J		Z		CC	
Race 6	X		T		M	
Race 7	R		E		N	
Race 8	O		K		S	
Race 9	C		AA		I	
Race 10	F		DD		A	
Round 5						
Race 1	C		R		L	
Race 2	K		A		G	
Race 3	H		J		F	
Race 4	S		I		W	
Race 5	E		P		CC	
Race 6	M		Y		V	
Race 7	AA		U		N	
Race 8	O		X		Z	
Race 9	T		BB		D	
Race 10	B		DD		Q	
Round 6						
Race 1	N		Q		S	
Race 2	AA		X		D	
Race 3	R		BB		M	
Race 4	H		C		O	
Race 5	V		F		Z	
Race 6	T		G		P	
Race 7	CC		I		K	
Race 8	DD		U		W	
Race 9	A		J		Y	
Race 10	B		L		E	

33 Car Round-Robin Schedule

	Lane 1		Lane 2		Lane 3	
	Car	Place	Car	Place	Car	Place

Round 1

	Car		Car		Car	
Race 1	A		B		C	
Race 2	D		E		F	
Race 3	G		H		I	
Race 4	J		K		L	
Race 5	M		N		O	
Race 6	P		Q		R	
Race 7	S		T		U	
Race 8	V		W		X	
Race 9	Y		Z		AA	
Race 10	BB		CC		DD	
Race 11	EE		FF		GG	

Round 2

	Car		Car		Car	
Race 1	N		V		Y	
Race 2	O		A		DD	
Race 3	Z		K		BB	
Race 4	L		FF		CC	
Race 5	B		U		AA	
Race 6	J		EE		C	
Race 7	W		GG		D	
Race 8	X		R		G	
Race 9	E		H		M	
Race 10	Q		F		S	
Race 11	I		P		T	

Round 3

	Car		Car		Car	
Race 1	C		Z		CC	
Race 2	D		DD		L	
Race 3	G		S		GG	
Race 4	M		BB		W	
Race 5	AA		T		FF	
Race 6	O		I		B	
Race 7	U		X		Y	
Race 8	N		P		A	
Race 9	H		F		EE	
Race 10	Q		E		J	
Race 11	V		R		K	

33 Car Round-Robin Schedule *continued*

	Lane 1		Lane 2		Lane 3	
	Car	Place	Car	Place	Car	Place
Round 4						
Race 1	FF		O		U	
Race 2	T		BB		EE	
Race 3	H		J		R	
Race 4	A		W		E	
Race 5	K		AA		M	
Race 6	L		G		B	
Race 7	CC		I		X	
Race 8	S		Y		P	
Race 9	C		N		Q	
Race 10	Z		V		D	
Race 11	DD		GG		F	
Round 5						
Race 1	W		J		FF	
Race 2	U		CC		Q	
Race 3	K		O		E	
Race 4	F		L		N	
Race 5	Y		A		T	
Race 6	GG		X		H	
Race 7	P		B		Z	
Race 8	AA		EE		G	
Race 9	I		C		S	
Race 10	R		D		BB	
Race 11	DD		M		V	
Round 6						
Race 1	GG		L		A	
Race 2	BB		G		J	
Race 3	R		C		O	
Race 4	T		D		N	
Race 5	X		S		Z	
Race 6	B		Q		V	
Race 7	CC		Y		W	
Race 8	EE		M		I	
Race 9	FF		DD		H	
Race 10	E		AA		P	
Race 11	F		U		K	

36 Car Round-Robin Schedule

	Lane 1 Car	Lane 1 Place	Lane 2 Car	Lane 2 Place	Lane 3 Car	Lane 3 Place
			Round 1			
Race 1	A		B		C	
Race 2	D		E		F	
Race 3	G		H		I	
Race 4	J		K		L	
Race 5	M		N		O	
Race 6	P		Q		R	
Race 7	S		T		U	
Race 8	V		W		X	
Race 9	Y		Z		AA	
Race 10	BB		CC		DD	
Race 11	EE		FF		GG	
Race 12	HH		II		JJ	
			Round 2			
Race 1	P		X		Y	
Race 2	Q		A		GG	
Race 3	BB		L		EE	
Race 4	M		HH		FF	
Race 5	B		W		AA	
Race 6	K		CC		II	
Race 7	Z		DD		D	
Race 8	JJ		T		H	
Race 9	C		I		E	
Race 10	S		F		N	
Race 11	G		J		O	
Race 12	R		U		V	
			Round 3			
Race 1	Z		BB		V	
Race 2	I		D		HH	
Race 3	X		U		J	
Race 4	O		P		CC	
Race 5	T		Y		E	
Race 6	GG		AA		DD	
Race 7	K		N		JJ	
Race 8	H		F		L	
Race 9	EE		G		B	
Race 10	C		M		Q	
Race 11	R		S		W	
Race 12	FF		II		A	

36 Car Round-Robin Schedule *continued*

	Lane 1		Lane 2		Lane 3	
	Car	Place	Car	Place	Car	Place
			Round 4			
Race 1	Q		V		CC	
Race 2	D		X		K	
Race 3	T		O		R	
Race 4	II		Y		BB	
Race 5	U		L		I	
Race 6	AA		E		H	
Race 7	N		DD		A	
Race 8	W		JJ		EE	
Race 9	J		M		B	
Race 10	FF		Z		C	
Race 11	F		G		P	
Race 12	GG		S		HH	
			Round 5			
Race 1	W		C		II	
Race 2	HH		H		N	
Race 3	CC		J		D	
Race 4	I		A		F	
Race 5	JJ		O		FF	
Race 6	U		B		Y	
Race 7	V		K		M	
Race 8	E		P		S	
Race 9	X		Q		G	
Race 10	AA		R		BB	
Race 11	DD		EE		T	
Race 12	L		GG		Z	
			Round 6			
Race 1	H		C		J	
Race 2	CC		EE		M	
Race 3	II		R		Z	
Race 4	Y		GG		G	
Race 5	DD		V		P	
Race 6	A		AA		K	
Race 7	L		FF		W	
Race 8	B		BB		Q	
Race 9	E		HH		U	
Race 10	F		JJ		X	
Race 11	N		D		T	
Race 12	O		I		S	

A 39 Car Round-Robin Schedule

	Lane 1		Lane 2		Lane 3	
	Car	Place	Car	Place	Car	Place

Round 1

Race 1	A		B		C	
Race 2	D		E		F	
Race 3	G		H		I	
Race 4	J		K		L	
Race 5	M		N		O	
Race 6	P		Q		R	
Race 7	S		T		U	
Race 8	V		W		X	
Race 9	Y		Z		AA	
Race 10	BB		CC		DD	
Race 11	EE		FF		GG	
Race 12	HH		II		JJ	
Race 13	KK		LL		MM	

Round 2

Race 1	B		Z		W	
Race 2	K		P		S	
Race 3	EE		U		Y	
Race 4	F		JJ		AA	
Race 5	BB		L		O	
Race 6	LL		H		E	
Race 7	I		CC		J	
Race 8	C		D		M	
Race 9	R		N		T	
Race 10	V		DD		FF	
Race 11	GG		MM		HH	
Race 12	X		II		Q	
Race 13	KK		A		G	

Round 3

Race 1	GG		JJ		Z	
Race 2	AA		MM		V	
Race 3	R		HH		EE	
Race 4	S		E		BB	
Race 5	Q		Y		CC	
Race 6	FF		KK		H	
Race 7	I		W		A	
Race 8	DD		F		M	
Race 9	II		O		P	
Race 10	G		LL		B	
Race 11	J		C		T	
Race 12	D		K		N	
Race 13	U		L		X	

	39 Car Round-Robin Schedule *continued*				
	Lane 1		Lane 2		Lane 3	
	Car	Place	Car	Place	Car	Place

Round 4

	Car	Place	Car	Place	Car	Place
Race 1	X		DD		Y	
Race 2	CC		EE		B	
Race 3	K		AA		W	
Race 4	N		P		C	
Race 5	L		M		Q	
Race 6	H		U		R	
Race 7	FF		D		S	
Race 8	HH		F		Z	
Race 9	O		GG		KK	
Race 10	T		G		BB	
Race 11	II		A		V	
Race 12	JJ		I		LL	
Race 13	MM		J		E	

Round 5

	Car	Place	Car	Place	Car	Place
Race 1	L		EE		II	
Race 2	W		Y		D	
Race 3	JJ		V		CC	
Race 4	LL		AA		A	
Race 5	DD		GG		G	
Race 6	Z		C		I	
Race 7	MM		M		U	
Race 8	B		Q		J	
Race 9	N		S		HH	
Race 10	T		O		F	
Race 11	E		R		FF	
Race 12	H		X		K	
Race 13	P		BB		KK	

Round 6

	Car	Place	Car	Place	Car	Place
Race 1	A		J		H	
Race 2	C		R		DD	
Race 3	CC		G		MM	
Race 4	E		I		GG	
Race 5	F		B		K	
Race 6	M		S		EE	
Race 7	O		FF		JJ	
Race 8	Q		V		D	
Race 9	AA		T		II	
Race 10	U		BB		LL	
Race 11	W		HH		L	
Race 12	Y		KK		N	
Race 13	Z		X		P	

Problems

1. **Scoring functions.** For the Pinewood Derby system described in this chapter, sketch scoring functions for the following Utilization of Resources Figures of Merit: Total Event Time, Number of Electrical Circuits, and Number of Adults.

2. **Matching functions.** This is a schedule for a nine car Pinewood Derby round robin.

9 Car Round-Robin Schedule

	Lane 1		Lane 2		Lane 3	
	Car	Place	Car	Place	Car	Place
			Round 1			
Race 1	A		B		C	
Race 2	D		E		F	
Race 3	G		H		I	
			Round 2			
Race 1	I		A		E	
Race 2	C		D		H	
Race 3	F		G		B	
			Round 3			
Race 1	H		F		A	
Race 2	B		I		D	
Race 3	E		C		G	
			Round 4			
Race 1	A		D		G	
Race 2	B		E		H	
Race 3	C		F		I	

With this schedule, each car races four times. Each scout races every other scout exactly once. Each car races in each lane at least once. Assume these are the race times for each car *not* in temporal order:

Car	Race Times, seconds			
A	2.40	2.41	2.42	2.43
B	2.41	2.42	2.43	2.44
C	2.42	2.43	2.44	2.45
D	2.43	2.44	2.45	2.46

Car	Race Times, seconds			
E	2.44	2.45	2.46	2.47
F	2.45	2.46	2.47	2.48
G	2.46	2.47	2.48	2.49
H	2.47	2.48	2.49	2.50
I	2.48	2.49	2.50	2.51

The system has three input ports, the three lanes. They accept data pairs as inputs, each data pair consisting of a car name and a time. The system has three output ports, the names of the first-, second-, and third-place cars. We will look only at the outputs at times $12n$, where $n = 0, 1, 2, 3, \dots$ We are judging this event on a basis of 1 point for first place, 2 points for second place, and 3 points for third place. At the end of four rounds, the car with the fewest total points wins. On the following pages we show three possible input trajectories, then several possible output trajectories. Your job is to derive a matching function that is appropriate for these trajectories.

Input Trajectory 1 (call it f1) with Output g1

Time		Lane 1 Car	Lane 1 Time	Lane 2 Car	Lane 2 Time	Lane 3 Car	Lane 3 Time	1st Place	2nd Place	3rd Place
					Round 1					
0	Race 1	A	2.40	B	2.41	C	2.42			
1	Race 2	D	2.43	E	2.44	F	2.45			
2	Race 3	G	2.46	H	2.47	I	2.48			
					Round 2					
3	Race 1	I	2.49	A	2.41	E	2.45			
4	Race 2	C	2.43	D	2.44	H	2.48			
5	Race 3	F	2.46	G	2.47	B	2.42			
					Round 3					
6	Race 1	H	2.49	F	2.47	A	2.42			
7	Race 2	B	2.43	I	2.50	D	2.45			
8	Race 3	E	2.46	C	2.44	G	2.48			
					Round 4					
9	Race 1	A	2.43	D	2.46	G	2.49			
10	Race 2	B	2.44	E	2.47	H	2.50			
11	Race 3	C	2.45	F	2.48	I	2.51			
12								A	B	C

Input Trajectory 2 (call it f2)

	Lane 1		Lane 2		Lane 3	
	Car	Time	Car	Time	Car	Time
			Round 1			
Race 1	A	**2.43**	B	2.41	C	2.42
Race 2	D	2.43	E	2.44	F	2.45
Race 3	G	2.46	H	2.47	I	2.48
			Round 2			
Race 1	I	2.49	A	2.41	E	2.45
Race 2	C	2.43	D	2.44	H	2.48
Race 3	F	2.46	G	2.47	B	2.42
			Round 3			
Race 1	H	2.49	F	2.47	A	2.42
Race 2	B	2.43	I	2.50	D	2.45
Race 3	E	2.46	C	2.44	G	2.48
			Round 4			
Race 1	A	**2.40**	D	2.46	G	2.49
Race 2	B	2.44	E	2.47	H	2.50
Race 3	C	2.45	F	2.48	I	2.51

Note: The differences between tables f2 and f1 are in boldface type.

Input Trajectory 3 (call it f3)

	Lane 1		Lane 2		Lane 3	
	Car	Time	Car	Time	Car	Time
			Round 1			
Race 1	A	**2.43**	B	**2.44**	C	2.42
Race 2	D	2.43	E	2.44	F	2.45
Race 3	G	2.46	H	2.47	I	2.48
			Round 2			
Race 1	I	2.49	A	2.41	E	2.45
Race 2	C	2.43	D	2.44	H	2.48
Race 3	F	2.46	G	2.47	B	2.42
			Round 3			
Race 1	H	2.49	F	2.47	A	2.42
Race 2	B	2.43	I	2.50	D	2.45
Race 3	E	2.46	C	2.44	G	2.48

	Lane 1		Lane 2		Lane 3	
	Car	Time	Car	Time	Car	Time
			Round 4			
Race 1	A	**2.40**	D	2.46	G	2.49
Race 2	B	**2.41**	E	2.47	H	2.50
Race 3	C	2.45	F	2.48	I	2.51

Input Trajectory 3 *continued*

Note: The differences between tables f3 and f1 are in boldface type.

Here are some possible values for the output trajectories at time $t = 12$.

$$g1(12) = (A, B, C),$$
$$g2(12) = (A, C, B),$$
$$g3(12) = (B, A, C),$$
$$g4(12) = (B, C, A),$$
$$g5(12) = (C, B, A),$$
$$g6(12) = (C, A, B),$$
$$g7(12) = (A, B, D),$$
$$g8(12) = (A, D, E).$$

For input trajectory f1, the total points are

$$A = 4$$
$$B = 5$$
$$C = 6$$
$$D = 7$$
$$E = 8$$
$$F = 9$$
$$G = 10$$
$$H = 11$$
$$I = 12$$

Therefore, an appropriate output is g1.

Now you should compute appropriate outputs for f2 and f3 and then write the matching function.

† How many input trajectories are possible? How many output trajectory values are possible for each time $12n$? How many matching functions are possible if you include all possible input and output trajectories? (Assume that the times given are only approximate and that electronic timing will

---

† This part of the problem is intended for students who have had a class in probability.

ensure that no race ends in a tie. During actual Pinewood Derbies with human judges there are ties and those races are rerun. Rerun races are very seldom ties. With electronic timing a whole derby is usually run with no ties.)

3. **Trade-off studies.** Assume that you get a new Grand Marshall for the Pinewood Derby who is not worried about irate parents. He says he will tell irate parents to "get lost," so he changes the weight on "Number of Irate Parents" to 0. Recalculate the final score for the three alternatives. Use the simulation data. (This is a long, tedious problem, but it will give you a good understanding of the trade-off process.)

4. † **Functional decomposition.** This question is seven pages long! The following is from the Pinewood Derby case study.

9 Car Round Robin Schedule

	Lane 1		Lane 2		Lane 3	
	Car	Place	Car	Place	Car	Place
			Round 1			
Race 1	A		B		C	
Race 2	D		E		F	
Race 3	G		H		I	
			Round 2			
Race 1	I		A		E	
Race 2	C		D		H	
Race 3	F		G		B	
			Round 3			
Race 1	H		F		A	
Race 2	B		I		D	
Race 3	E		C		G	
			Round 4			
Race 1	A		D		G	
Race 2	B		E		H	
Race 3	C		F		I	

With this schedule, each car races four times. Each scout races every other scout exactly once. Each car races in each lane at least once. Assume these are the finish times *not* in temporal order.

---

† This problem uses more detailed notation than is used in the text.

Car	Race Times, seconds			
A	2.40	2.41	2.42	2.43
B	2.41	2.42	2.43	2.44
C	2.42	2.43	2.44	2.45
D	2.43	2.44	2.45	2.46
E	2.44	2.45	2.46	2.47
F	2.45	2.46	2.47	2.48
G	2.46	2.47	2.48	2.49
H	2.47	2.48	·2.49	2.50
I	2.48	2.49	2.50	2.51

Our input/output requirement has three input ports, the three lanes. They accept data pairs as inputs, each data pair consisting of a car name (A through I) and a finish time (from 2.40 to 2.51). The system has three output ports that present the names of the first, second, and third place cars. We are judging this event on a basis of 1 point for first place, 2 points for second place, and 3 points for third place. At the end of four rounds, the car with the fewest total points wins. On the following pages we show three possible input trajectories, then several possible output trajectories.

Input Trajectory 1 (call it f1) with Output g1

		Inputs					Outputs		
		Lane 1		Lane 2		Lane 3		1st	2nd	3rd
Time		Car	Time	Car	Time	Car	Time	Place	Place	Place
				Round 1						
0	Race 1	A	2.40	B	2.41	C	2.42			
1	Race 2	D	2.43	E	2.44	F	2.45			
2	Race 3	G	2.46	H	2.47	I	2.48			
				Round 2						
3	Race 1	I	2.49	A	2.41	E	2.45			
4	Race 2	C	2.43	D	2.44	H	2.48			
5	Race 3	F	2.46	G	2.47	B	2.42			
				Round 3						
6	Race 1	H	2.49	F	2.47	A	2.42			
7	Race 2	B	2.43	I	2.50	D	2.45			
8	Race 3	E	2.46	C	2.44	G	2.48			
				Round 4						
9	Race 1	A	2.43	D	2.46	G	2.49			
10	Race 2	B	2.44	E	2.47	H	2.50			
11	Race 3	C	2.45	F	2.48	I	2.51			
12								A	B	C

Input Trajectory 2 (call it f2)

	Lane 1		Lane 2		Lane 3	
	Car	Time	Car	Time	Car	Time
Round 1						
Race 1	A	**2.43**	B	2.41	C	2.42
Race 2	D	2.43	E	2.44	F	2.45
Race 3	G	2.46	H	2.47	I	2.48
Round 2						
Race 1	I	2.49	A	2.41	E	2.45
Race 2	C	2.43	D	2.44	H	2.48
Race 3	F	2.46	G	2.47	B	2.42
Round 3						
Race 1	H	2.49	F	2.47	A	2.42
Race 2	B	2.43	I	2.50	D	2.45
Race 3	E	2.46	C	2.44	G	2.48
Round 4						
Race 1	A	**2.40**	D	2.46	G	2.49
Race 2	B	2.44	E	2.47	H	2.50
Race 3	C	2.45	F	2.48	I	2.51

Note: The differences between tables f2 and f1 are in boldface type.

Input Trajectory 3 (call it f3)

	Lane 1		Lane 2		Lane 3	
	Car	Time	Car	Time	Car	Time
Round 1						
Race 1	A	**2.43**	B	**2.44**	C	2.42
Race 2	D	2.43	E	2.44	F	2.45
Race 3	G	2.46	H	2.47	I	2.48
Round 2						
Race 1	I	2.49	A	2.41	E	2.45
Race 2	C	2.43	D	2.44	H	2.48
Race 3	F	2.46	G	2.47	B	2.42
Round 3						
Race 1	H	2.49	F	2.47	A	2.42
Race 2	B	2.43	I	2.50	D	2.45
Race 3	E	2.46	C	2.44	G	2.48

		Input Trajectory 3 *continued*			
	Lane 1		Lane 2		Lane 3
	Car	Time	Car	Time	Car	Time
		Round 4			
Race 1	A	**2.40**	D	2.46	G	2.49
Race 2	B	**2.41**	E	2.47	H	2.50
Race 3	C	2.45	F	2.48	I	2.51

Note: The differences between tables f3 and f1 are in boldface type.

We now show values of some possible ouputs. (*Note:* these are not technically trajectories, but they are only values of trajectories for some particular time.)

$$g1 = (A, B, C),$$
$$g2 = (A, C, B),$$
$$g3 = (B, A, C),$$
$$g4 = (B, C, A),$$
$$g5 = (C, B, A),$$
$$g6 = (C, A, B),$$
$$g7 = (A, B, D),$$
$$g8 = (A, D, E),$$
$$g9 = (J, J, J),$$
$$g10 = (A, B, E).$$

For simplicity, assume that no individual race ends in a tie. During actual Pinewood Derbies with human judges there are ties and those races are rerun. The rerun races are very seldom ties. With electronic timers, a whole derby is usually run with no ties.

The following is a set theoretic description of what we have just said in words. First we give the original Input/Output and Functional Requirement for the Pinewood Derby Part 0 (IORpwd0). Later we do the same for Parts 1, 2, and 3 (IORpwd1, etc).

```
IORpwd0 = (TRpwd0, IRpwd0, ITRpwd0, ORpwd0,
           OTRpwd0, MRpwd0),
```

where

```
TRpwd0 = IJS[0-12],
/*These requirements must be satisfied for the
   times 0 to 12.*/
IRpwd0 = IR1pwd0 × IR2pwd0 × IR3pwd0,
   IR1pwd0 = (ALPHABET[A-I], RLS[2.40-2.51])
      /*Name of car and finish time for lane 1*/
```

```
/*The notation ALPHABET[A-I] means any letter*/
/*of the alphabet between A and I*/
  IR2pwd0 = (ALPHABET[A-I], RLS[2.40-2.51])
    /*Name of car and finish time for lane 2*/
  IR3pwd0 = (ALPHABET[A-I], RLS[2.40-2.51])
    /*Name of car and finish time for lane 3*/
ITRpwd0 = FNS(TRpwd0, IRpwd0),
ORpwd0 = OR1pwd0 × OR2pwd0 × OR3pwd0,
  OR1pwd0 = ALPHABET[A-I]
    /*Name of first place car*/
  OR2pwd0 = ALPHABET[A-I]
    /*Name of second place car*/
  OR3pwd0 = ALPHABET[A-I]
    /*Name of third place car*/
OTRpwd0 = FNS(TRpwd0, ORpwd0),
    /*Any trajectories that can be made with the
      above input and output requirements are
      legal.*/
/*The following line says that MRpwd0 is a
  function of f and G: where f is an element of
  the set ITRpwd0; and G is a subset of the set
  OTRpwd0; and G is further restricted in that
  the elements of G, represented with g, are
  elements of the set OTRpwd0;*/
MRpwd0 = {(f, G): f ∈ ITRpwd0; G is a subset of
            OTRpwd0; G = {g: g ∈ OTRpwd0;
  if (f = f1) then g(12) = g1;
  else if (f = f2) then g(12) = g4;
  else if (f = f3) then g(12) = g6}}.
```

Now your engineers come to you and say, "It's going to be hard to build a system that satisfies IORpwd0, but in the back room we have systems on the shelf that satisfy IORpwd1, IORpwd2, and IORpwd3." They also claim that ICRpwd4 (which produces IORpwd4) decomposes IORpwd0 into IORpwd1, IORpwd2, and IORpwd3. Do you believe them? Draw or state what ICRpwd4 must be. Define the relationships between OTRpwd0 and OTRpwd4 and between MRpwd0 and MRpwd4. If you implement the system using the three systems your engineers recommend, what aspects of the customers requirements as stated in IORpwd() will not be satisfied? Are there any new features the customer did not request?

```
IORpwd1 = (TRpwd1, IRpwd1, ITRpwd1, ORpwd1,
            OTRpwd1, MRpwd1),
```

where

```
TRpwd1 = IJS[0-12],
```

```
IRpwd1 = IR1pwd1 × IR2pwd1 × IR3pwd1,
  IR1pwd1 = (ALPHABET[A-J], RLS[2.40-2.51])
    /*Name of car and finish time for lane 1*/
  IR2pwd1 = (ALPHABET[A-J], RLS[2.40-2.51])
    /*Name of car and finish time for lane 2*/
  IR3pwd1 = (ALPHABET[A-J], RLS[2.40-2.51])
    /*Name of car and finish time for lane 3*/
ITRpwd1 = FNS(TRpwd1, IRpwd1),
ORpwd1 = ALPHABET{A-I] /*Name of first place
                               car*/
OTRpwd1 = FNS(TRpwd1, ORpwd1),
MRpwd1 = {(f, G): where f ∈ ITRpwd1; G is a
            subset of OTRpwd1;
            G = {g: g ∈ OTRpwd1; n ∈ IJS[0-11];
                g(n) = g9
                if (f = f1) then g(12) = A;
                else if (f = f2) then g(12) = B;
                else if (f = f3) then g(12) = C;
                else g(12) = g9}}.
IORpwd2 = (TRpwd2, IRpwd2, ITRpwd2, ORpwd2,
            OTRpwd2, MRpwd2),
```

where

```
TRpwd2 = IJS[0-12],
IRpwd2 = IR1pwd2 × IR2pwd2 × IR3pwd2,
  IR1pwd2 = (ALPHABET[A-J], RLS[2.40-2.51])
    /*Name of car and finish time for lane 1*/
  IR2pwd2 = (ALPHABET[A-J], RLS[2.40-2.51])
    /*Name of car and finish time for lane 2*/
  IR3pwd2 = (ALPHABET[A-J], RLS[2.40-2.51])
    /*Name of car and finish time for lane 3*/
ITRpwd2 = FNS(TRpwd2, IRpwd2),
ORpwd2 = ALPHABET[A-I] /*Name of second place
                               car*/
OTRpwd2 = FNS(TRpwd2, ORpwd2),
MRpwd2 = {(f, G): where f ∈ ITRpwd2; G is a
            subset of OTRpwd2;
            G = {g: g ∈ OTRpwd2; n ∈ IJS[0-11];
                g(n) = g9
                if (f = f1) then g(12) = B;
                else if (f = f2) then g(12) = C;
                else if (f = f3) then g(12) = A;
                else g(12) = g9}}.
IORpwd3 = (TRpwd3, IRpwd3, ITRpwd3, ORpwd3,
            OTRpwd3, MRpwd3),
```

where

```
TRpwd3 = IJS[0-12],
IRpwd3 = IR1pwd3 × IR2pwd3 × IR2pwd3,
  IR1pwd3 = (ALPHABET[A-J], RLS[2.40-2.51])
    /*Name of car and finish time for lane 1*/
  IR2pwd3 = (ALPHABET[A-J], RLS[2.40-2.51])
    /*Name of car and finish time for lane 2*/
  IR3pwd3 = (ALPHABET[A-J], RLS[2.40-2.51])
    /*Name of car and finish time for lane 3*/
ITRpwd3 = FNS(TRpwd3, IRpwd3),
ORpwd3 = ALPHABET[A-I] /*Name of third place
                            car*/
OTRpwd3 = FNS(TRpwd3, ORpwd3),
MRpwd3 = {(f, G); where f ∈ ITRpwd3; G is a
           subset of OTRpwd3;
           G = {g: g ∈ OTRpwd3; n ∈ IJS[0-11];
               g(n) = g9
               if (f = f1) then g(12) = C;
               else if (f = f2) then g(12) = A;
               else if (f = f3) then g(12) = B;
               else g(12) = g9}}.
```

chapter six

SIERRA

SIERRA (Systems and Industrial Engineering Railroad Assignment) is an undergraduate project that teaches students to think in a systematic way. Though the particular documentation shown here is not used by a company, it is representative of the type of documentation used in industry. All the information needed to design the system is given, along with a recommended concept for full-scale engineering. The documentation presented in this chapter has not been continuously polished until it is perfect, but rather it represents how documentation may look for a genuine project. Added to the documents are boxes in italics (such as this one) with additional comments or explanations that would not normally appear in a systems engineering report. Some of these are comments on the quality of the system and how it, or the documentation, may or may not be improved, and some are general statements that help to clarify the process for engineering students.

Contents

6.1 Document 1: Problem Situation

> The Problem Situation document is the executive summary. It explains the problem that needs to be solved. It is written in plain language and is intended for management.

6.1.1 The top level system function

The top level system function is to control two model trains in the undergraduate SIE-370 lab in such a way as to avoid collisions and to maximize the number of trips completed by each train during the test period. The student-built controllers will demonstrate the students' success in designing a system and will also serve as a demonstration for future visitors to the department.

6.1.2 History of the problem and the present system

Since 1985, the students of SIE-370 have used a pair of trains (see Figure 6.1) to learn more about computers and systems engineering. The train tracks have

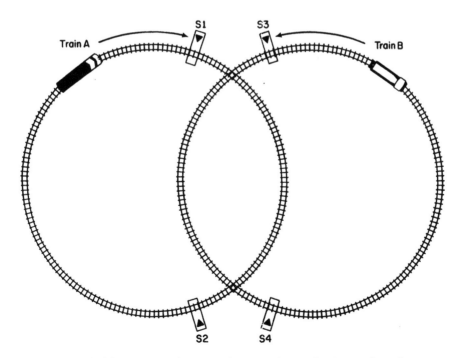

Figure 6.1 Model train setup showing relative positions of trains, track, and sensors.

two crossover points where the trains can collide. The students' task is to build a system that prevents collisions and maximizes the number of trips completed by each train. Students must use existing detection and power control devices. This project is known as the Systems and Industrial Engineering Railroad Assignment (SIERRA).

6.1.3   The customer

6.1.3.1   Owners
The system will be owned by the SIE Department of the University of Arizona.

6.1.3.2   Bill payers: The client
Costs of the system are paid through the SIE departmental budget. Students will pay for any hardware or software they decide to use that is not available in the department.

6.1.3.3   Users
The system will be used by the teaching assistants (TAs) in the SIE-370 class to verify the students' design, and it will be used by professors and TAs to demonstrate the SIE systems engineering philosophy to department visitors.

6.1.3.4   Operators
The system will be operated by the students who built the system and by the TAs in the SIE-370 lab.

6.1.3.5   Beneficiaries
The students are the beneficiaries of the system in that they gain knowledge and experience. They learn good systems engineering documentation practices and techniques.

6.1.3.6   Victims
Victims of the system are those who feel that the system had a negative impact on them. The only known victims of SIERRA are students who complain that their grades do not reflect the energy they expended on the project.

If systems engineers say that there are no victims, they probably have not thought about the problem. One of the reasons for the rigid outline of these documents is to force the systems engineers to think about such possible problems. Dr. Bahill usually gives failing marks to all projects that say, "There are no known victims of the system."

6.1.3.7 Technical representatives to systems engineering
Dr. Bahill, Professor of Systems and Industrial Engineering at the University of Arizona, and the SIE-370 TAs will provide the technical interface for the SIE-370 class.

6.1.4 Technical personnel and facilities

6.1.4.1 Life Cycle Phase 1: Requirements development
Dr. Bahill and Bill Karnavas are the technical interface to the Systems Engineering Department throughout Phase 1. All requirements data will be supplied by them. Systems engineering personnel will provide their own supplies and computer equipment throughout this phase.

6.1.4.2 Life Cycle Phase 2: Concept development
The SIE-370 lab and all its facilities are available for Phase 2. The students will be the concept developers, with help from Dr. Bahill and the TA in charge of the lab.

6.1.4.3 Life Cycle Phase 3: Full-scale engineering development
Full-scale engineering development will be done in the lab or at home by the students. Help from Dr. Bahill and the TA in charge of the lab will be provided as needed.

6.1.4.4 Life Cycle Phase 4: System development
System development will be performed by the students in the SIE-370 lab or at home. Help from Dr. Bahill and the TA in charge of the lab will be provided as needed.

6.1.4.5 Life Cycle Phase 5: System test and integration
The system test will be performed by the TA in charge of the SIE-370 lab. The students will be required to integrate their designs with the existing test equipment, as detailed in the requirements. The tests will be conducted in the SIE-370 lab.

6.1.4.6 Life Cycle Phase 6: Operations support and modification
Since the life of the operating system is only 30 minutes, there is no support or modification effort. The system must pass the system test and integration phase to be considered operational. Only rugged systems will be considered for retention as demonstration systems for the future.

6.1.4.7 Life Cycle Phase 7: Retirement and replacement
The system is retired after a successful test and acceptance. The permanent demonstration systems may be replaced in the following year when the next class does the project. Students must disassemble their prototypes and return the components to the SIE lab.

6.1.5 System environment

6.1.5.1 Social impact

In interviews with alumni of the past thirty years Dr. Bahill always asks, "Of all the tools and techniques that we have taught you, which have you found to be the most valuable?" The most common answers have been

1. the principles of system design,
2. learning to work with other people on a project, and
3. learning to write and present a systems engineering report.

The social impact of this project is the value in learning those three lessons.

6.1.5.2 Economic impact

The laboratory for this course is the most expensive one in the SIE Department. Dr. Bahill must continually compete with other professors for hardware and personnel resources.

6.1.5.3 Environmental impact

If the necessary hardware were affordable, the students would be permitted to test their circuits at home, thus reducing the occupancy of the SIE-370 laboratory and its associated maintenance. This would also reduce the transportation load caused by students entering and leaving the lab.

The environment of the laboratory is affected by students breaking parts and equipment and leaving trash behind.

6.1.5.4 Interoperability

The system must interface with the existing train power controls and train location monitors. The sensors are switches S1, S2, S3, and S4; they are located as shown in Figure 6.1.

6.1.6 Systems engineering management plan

Systems engineers will design the project using the seven systems engineering documents. These documents will be continually updated as the design progresses using the software package Systems Engineering Design Software (SEDSO). In addition, the students will be responsible for the project throughout the life cycle and will supply any additional data the instructor needs to determine whether the full-scale engineering project was completed correctly.

6.2 Document 2: Operational Need

> The Operational Need Document is a detailed description of the prob-
> lem in plain language. It is intended for management, the customer,
> and systems engineers. It contains some of the same information
> that is in Document 1, only it is more detailed.

6.2.1 Deficiency

Two HO-gauge model trains are on circular tracks that intersect at two points;
the trains can collide at the intersections.

6.2.2 Input/Output and Functional Requirement

6.2.2.1 Time scale

The system shall have a time scale resolution of milliseconds. The life of the
system will be 30 minutes. A few superior systems will be kept for years to
serve as demonstrations for Systems Engineering Department visitors.

6.2.2.2 Inputs

The system will have four inputs that indicate train position (see Figure 6.1
in Document 1); they are labeled S1, S2, S3, and S4.

6.2.2.3 Input trajectories

The input trajectories will be restricted as follows:

1. Train A will activate switch S1 and then, after an indeterminate
   amount of time, switch S2. The train's length is such that at no time
   will S1 and S2 both be activated.

2. Train B will activate switch S3 and then, after an indeterminate
   amount of time, switch S4. The train's length is such that at no time
   will S3 and S4 both be activated.

3. We can safely assume that at no time will S2 and S4 both be activated,
   because for this to occur, both trains would have to be in the danger
   zone at the same time, which is not permitted.

4. It is possible, as a result of switch bouncing, that a switch will read
   ON, then OFF, then ON again in rapid succession. If this occurs within
   10 milliseconds, it should be considered as a steady ON signal.

6.2.2.4 Outputs

The outputs are power ON or OFF for each train (PA and PB).

6.2.2.5 Output trajectories

The output trajectories will be restricted as follows: Power to one or both of the trains must be ON at all times. We assume that power will be OFF or ON 1 millisecond after the output is activated. Since a train's momentum may cause it to travel as much as 1 foot after the power is turned OFF, a maximum safe train speed specification is required for each design. Power to one train can be turned OFF to prevent collisions, then turned back ON again when it is safe.

6.2.2.6 Matching function

The required matching between input trajectories and output trajectories is as follows: After S2 or S4 is activated, power is turned ON to both trains.

> *Of course, a more restrictive matching function could have been written, but it is often easiest to at least start out with a simple matching function.*

6.2.3 Technology Requirement

6.2.3.1 Available money

Computer time and student labor and engineering time are free. Also, components available within the lab are free.

6.2.3.2 Available time

Students have 1 month to complete the project.

6.2.3.3 Available components

Integrated circuits (TTL type) available in the lab are sufficient for building a system. Students who decide to use other components must purchase them at their own expense.

A Motorola computer (with a 68000 microprocessor) that can be interfaced to the inputs and outputs is available. Software developed on other systems must be downloaded to this computer.

A software development environment is provided. It allows students to use either Unix-based PCs or DOS-based IBM PCs.

6.2.3.4 Available technologies

Three technologies must be used on this project:

1. Components: Students must design and build a controller using TTL integrated circuits, resistors, capacitors, LEDs, wires, a power supply, and a protoboard.

2. Assembly language programming: Students must write an assembly language program and download it to the Motorola 68000 computer interfaced to the input and output devices.

3. High-level programming: Students must write a program in a high-level language (e.g., C, FORTRAN, Pascal) and download it to the Motorola 68000 computer interfaced to the input and output devices.

6.2.3.5 Required interfaces
All units are required to interface to the following:

1. Four location-detector switches: The four sensors—S1, S2, S3, and S4—are available as bits 1, 2, 3, and 4, respectively, of the microcomputer word at location $10010 (the symbol $ indicates that this is a binary number). This is a parallel port on the lab's Motorola controller.

2. Power controllers for Train A and Train B: The two outputs (PA and PB) are available as bits 5 and 6 of the same microcomputer word at address $10010.

3. Six connection points for attaching the test equipment: These will be positioned at the end of any hardware board in the following sequence: S1, S2, S3, S4, GND, GND, PA, and PB, where GND stands for electrical ground. A cable with an eight-pin plug will be plugged into this section of the hardware board.

6.2.3.6 Standards, specifications, and other restrictions
Students are expected to follow all guidelines for safety and manufacturing outlined by the TA of the lab.

6.2.4 Input/Output Performance Requirement

1. *Number of Collisions:* The total number of times the two trains come into physical contact.

2. *Trips by Train A:* The total number of completed trips for Train A.

3. *Trips by Train B:* The total number of completed trips for Train B.

4. *Spurious Stops by A:* The total number of spurious stops by Train A. A spurious stop is one that is not needed to avoid a collision.

5. *Spurious Stops by B:* Total number of spurious stops by Train B. A spurious stop is one that is not needed to avoid a collision.

6. *Availability:* If not in failure mode, the system will be considered available when it is submitted to the TA for the testing period. The system is in failure mode if it does not interface correctly to the input detectors or if it does not control the outputs to both trains.

7. *Reliability:* The system will be judged reliable if it does not enter failure mode while it is being tested. The system is in failure mode if it does not interface correctly to the input detectors or if it does not control the outputs to both trains.

Input/Output Performance Requirements and Utilization of Resources Requirements can be thought of as the measurements necessary to ensure that the requirements are satisfied. These are also indicators of the quality of the system.

6.2.5 *Utilization of Resources Requirement*

1. *Completion Time:* The number of hours the project was completed before the due date.

2. *Acquisition Cost:* The total cost of creating the unit. This is a combination of the next three requirements.

    2.1. *Cost, Quantity 1:* The cost to create 1 unit.

    2.2. *Cost, Quantity 1000:* The estimated cost per unit to create 1000 units.

    2.3. *Cost, Quantity 1,000,000:* The estimated cost per unit to create 1,000,000 units.

6.2.6 *Trade-Off Requirement*

The trade-off analysis will give equal weight to the performance and resource requirements.

6.2.7 *System Test Requirement*

The student's system will be tested for a total of 10 minutes. Each system will be subjected to five separate tests, each having a two minute duration. The initial positions of the trains and their speeds will be varied as defined in Document 3.

1. Test 1 will determine if Train B stops as it enters the collision area when Train A is already there.

2. Test 2 will determine if Train A stops as it enters the collision area when Train B is already there.

3. Test 3 will determine if collisions are avoided when A enters the danger area and B then tries to enter, followed by A leaving and B then entering, followed by B leaving and going around the entire track and re-entering the danger area before A tries to enter again.

4. Test 4 will determine if collisions are avoided when B enters the danger area and A then tries to enter, followed by B leaving and A then entering, followed by A leaving and going around the entire track and re-entering the danger area before B tries to enter again.

5. Test 5 will determine what happens when both A and B enter the danger area (and activate S1 and S3, respectively) simultaneously.

A stipulation of simultaneous changes on S1 and S3 is not good for a test because of the difficulty in defining "simultaneous." For example, a human considers two events occurring within one millisecond to be simultaneous and a computer with a 1 MHz clock requires them to occur within one microsecond, whereas the TTL circuitry defines simultaneous events as those that occur within nanoseconds of each other.

The system will satisfy the observance criteria if

1. all performance and resource requirements are observed as described in Document 3.

The system will be in compliance if

1. all performance requirements are within the upper and lower thresholds and
2. all resource requirements are within the upper and lower thresholds.

The system will be acceptable if

1. it satisfies the observance requirement,
2. it is in compliance, and
3. the trains do not collide during any test.

The built and tested system will be in conformance if

1. it is acceptable and
2. it satisfies the customer.

6.2.8 Rationale for operational need

The data and specifications were provided during an SIE-650 class and in subsequent discussions with Dr. Bahill.

6.3 Document 3: System Requirements

The Systems Requirements Document is a succinct mathematical description of the Input/Output and Functional Requirements, Technology Requirements, and Test Requirements as described in Document 2. Its audience is systems engineers. Any modeling language can be used for this document. In the SIERRA case study we use our set theoretic notation.

6.3.1 The system requirement

The System Design Problem entails stating the following requirements:

- Input/Output and Functional Requirement,
- Technology Requirement,
- Input/Output Performance Requirement,
- Utilization of Resources Requirement,
- Trade-Off Requirement, and
- System Test Requirement.

Each of these requirements is mathematically stated in the following sections.

6.3.2 Input/Output and Functional Requirement

6.3.2.1 Time scale
`TRP1` is the time scale of SIERRA expressed in milliseconds. The life expectancy of the system is 30 minutes. This becomes 30 minutes × 60 seconds/minute × 1000 milliseconds/second = 1,800,000.

```
TRP1 = IJS [0, 1800000].
```

6.3.2.2 Inputs
`IRP1` represents the set of system inputs for SIERRA. These are the switches that detect the location of the trains.

```
IRP1 = I1P1 × I2P1 × I3P1 × I4P1
```

```
I1P1 = {0,1} /*where Input Port 1=S1 represents
                  the condition of switch 1.
                  Train A not present = 0,
                  present = 1*/
I2P1 = {0,1} /*where Input Port 2=S2 represents
                  the condition of switch 2.
                  Train A not present = 0,
                  present = 1*/
I3P1 = {0,1} /*where Input Port 3=S3 represents
                  the condition of switch 3.
                  Train B not present = 0,
                  present = 1*/
I4P1 = {0,1} /*where Input Port 4=S4 represents
                  the condition of switch 4.
                  Train B not present = 0,
                  present = 1*/
```

6.3.2.3 *Input trajectories*

ITRP1 is the set of input trajectories. It is the set of all possible inputs (IRP1) over the time scale (TRP1). Formally,

```
ITRP1 = {f: f ∈ FNS(TRP1,IRP1);
           for every t ∈ TRP1,
             Let p1 be the value of I1P1 for a
                given t,
             let p2 be the value of I2P1 for a
                given t,
             let p3 be the value of I3P1 for a
                given t, and
             let p4 be the value of I4P1 for a
                given t,
           then (p1 = 1 and p2 = 1) = {} and
              (p3 = 1 and p4 = 1) = {} and
              (p2 = 1 and p4 = 1) = {}}
```

Interpreting this notation: Train A cannot be at Switches 1 and 2 at the same time (it is not long enough); Train B cannot be at Switches 3 and 4 at the same time (it is not long enough either); and Train A cannot be at Switch 2 (leaving the danger zone) at the same time Train B is at Switch 4 (leaving the danger zone) because they would have already collided.

This description ignores the possibility of noise on input lines. Some student designs did consider noise; these designs were more robust.

6.3.2.4 *Outputs*

ORP1 represents the set of outputs.

```
ORP1 = 01P1 × 02P1
```

where

```
01P1 = {0,1} /*where Output Port 1=PA represents
                the power to Train A.
                Power to Train A: OFF = 0,
                ON = 1*/
02P1 = {0,1} /*where Output Port 2=PB represents
                the power to Train B.
                Power to Train B: OFF = 0,
                ON = 1*/
```

6.3.2.5 *Output trajectories*

OTRP1 is the set of all output trajectories for SIERRA. OTRP1 is the set of all possible outputs (ORP1) over the time scale (TRP1), with the exception that at no time can the power be OFF to both trains. Formally,

```
OTRP1 = {f: f ∈ FNS(TRP1,ORP1);
            for every t ∈ TRP1,
              let q1 be the value of 01P1 for a
              given t,
              let q2 be the value of 02P1 for a
              given t,
            then if (q1 = 0) then q2 = 1 and
            if (q2 = 0) then q1 = 1}
```

Interpretation: For every time and output combination, if Port 1—the power to Train A—is OFF, then Port 2—the power to Train B—is ON, and vice versa.

6.3.2.6 *Matching function*

MRP1 is the matching function.

```
MRP1 = {(f,G): p ∈ ITRP1; G ∈ OTRP1;
           G={q: q ∈ ORP1; t ∈ TRP1 then
              if p(t)=(0,1,0,0) or p(t)=(0,0,0,1)),
              then q(t+1)=(1,1)}}
```

Interpretation: For any input and output over the time scale, if Switch 2 is ON or Switch 4 is ON, then the next output puts the power ON for both trains. We have matched the inputs and outputs by definition.

Once again we note that more complicated matching functions are possible. This reflects the simple one presented in Section 6.2.2.6.

6.3.3 Technology Requirement

6.3.3.1 Available money

Computer time on the Unix-based PC, the DOS-based IBM PC, and the Motorola controller is free. Student engineering time is also free. Components available in the lab are free to students, but students must purchase any other components they decide to use.

6.3.3.2 Available time

Students have 1 month to complete the project.

6.3.3.3 Available components

The software-based Motorola controller must be used for the train controllers.

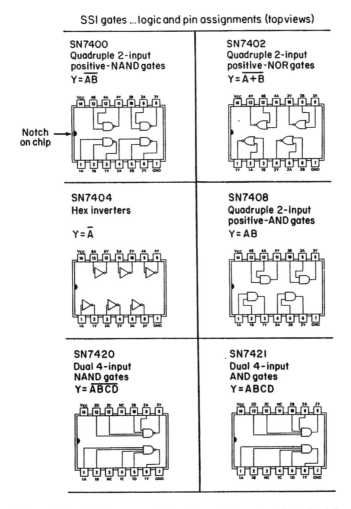

Figure 6.2a TTL integrated circuits readily available in the laboratory.

The Unix-based PC or the DOS-based IBM PC in the lab may be used to write the software for the Motorola controller.

> *The Motorola controller is actually an MC68000 Educational Computer Board (ECB). It is a single printed circuit card containing (1) a microprocessor, (2) 32 kilobytes of memory, (3) an operating system loaded in a ROM, and (4) parallel and serial input-output ports. Computers that are specialized for controlling devices are often called controllers.*

Components are available in the SIE-370 lab for the protoboard controller. Figures 6.2*a* and 6.2*b* show the TTL integrated circuits that are available.

6.3.3.4 *Available technologies*

1. Components: Students must create a circuit that is able to detect the inputs, make the correct decisions, and control the outputs.

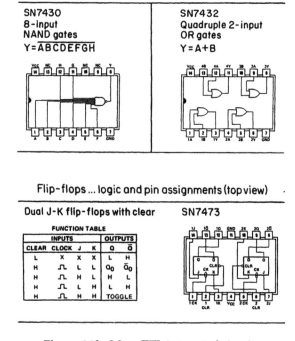

Figure 6.2*b* More TTL integrated circuits.

2. Assembly language programming: Students must write an assembly language program and download it to the Motorola controller interfaced to the input and output devices.

3. High-level programming: Students must write a program in a high-level computer language (e.g., C, Pascal, FORTRAN) and download it to the Motorola controller interfaced to the input and output devices.

6.3.3.5 Required interfaces

All units are required to interface to the following:

1. Four location-detector switches: The four sensors—S1, S2, S3, and S4—are available as bits 1, 2, 3, and 4, respectively, of the microcomputer word at location $10010. This is a parallel port on the lab's Motorola controller.

2. Power controllers for Train A and Train B: The two outputs (PA and PB) are available as bits 5 and 6 of the same microcomputer word at address $10010.

3. Six connection points for attaching the test equipment: These will be positioned at the end of any hardware board in the following sequence: S1, S2, S3, S4, GND, GND, PA, and PB, where GND stands for electrical ground. A cable with an eight-pin plug will be plugged into this section of the hardware board.

These paragraphs are identical to those of Section 6.2.3.5. This is why we used a hypertext tool to prepare these documents. It is important to note that many mistakes made during system integration are caused by poorly defined interfaces.

6.3.3.6 Form, fit, and other restrictions

The hardware controller must be built on a protoboard with an area of less than 30 square inches. The power lines are +5 VDC and ground. The controller must attach to the train conditioning circuits by means of the plug described in Section 6.3.3.5. The assembly language and higher-level language programs must be stored on a floppy diskette readable by either the Unix-based PC or the DOS-based IBM PC.

6.3.3.7 Standards and specifications

The student must use safe engineering practices when working with electrical power as per the SIE-370 laboratory and Arizona restrictions.

6.3.4   *Input/Output Performance Requirement*

6.3.4.1   *Definition of Performance Figures of Merit*
The overall performance figure of merit is denoted IFOP1 and is computed as follows:

$$\text{IFOP1} = \text{ISF1P1} * \text{IW1P1} + \text{ISF2P1} * \text{IW2P1} + \ldots + \text{ISF}n\text{P1} * \text{IW}n\text{P1}$$

where n is the total number of I/O Performance Figures of Merit and

$$\text{ISF}i\text{P1} = \text{IS}i\text{P1}(\text{IF}i\text{P1}(\text{FSD})) \text{ for } i = 1 \text{ to } n$$

6.3.4.2   *Lower, upper, baseline, and scoring parameters*
In this section, the following naming convention for variables is used:

$\quad$ IFiP1 = the ith figure of merit measured per the test plan,
$\quad$ IBiP1 = the baseline value for the ith figure of merit,
$\quad$ ILTHiP1 = lower threshold for the ith figure of merit,
$\quad$ ISFiP1 = score for the ith figure of merit,
$\quad$ ISiP1 = scoring function for the ith figure of merit,
$\quad$ ISLiP1 = slope for the ith figure of merit,
$\quad$ IUTHiP1 = upper threshold for the ith figure of merit,
$\quad$ IWiP1 = weight for the ith figure of merit, and
$\quad$ SSF = standard scoring function.

These parameters are created by systems engineering to describe the value of the measurements. For example, the number of collisions has a lower limit of 0 and no upper limit. The baseline, or expected, value is 0.6. The baseline parameter indicates the figure of merit that will yield a value of 0.6 on a scale from 0 to 1. The slope is measured at the baseline; it shows how quickly the scaled value changes at that point.

1. Number of Collisions

Score	IS1P1 = SSF(ILTH1P1,IB1P1,ISL1P1,IJS++)
Lower Threshold	ILTH1P1 = 0
Baseline	IB1P1 = 0.6
Upper Threshold	IUTH1P1 = ∞
Slope	ISL1P1 = –2

   IJS++ indicates all positive integers.

2. Trips by Train A

Score	IS2P1 = SSF(ILTH2P1,IB2P1,ISL2P1,IJS++)
Lower Threshold	ILTH2P1 = 0
Baseline	IB2P1 = 5
Upper Threshold	IUTH2P1 = ∞
Slope	ISL2P1 = 0.2

3. Trips by Train B

Score	IS3P1 = SSF(ILTH3P1,IB3P1,ISL3P1,IJS++)
Lower Threshold	ILTH3P1 = 0
Baseline	IB3P1 = 5
Upper Threshold	IUTH3P1 = ∞
Slope	ISL3P1 = 0.2

4. Spurious Stops by A

Score	IS4P1 = SSF(ILTH4P1,IB4P1,ISL4P1,IJS++)
Lower Threshold	ILTH4P1 = 0
Baseline	IB4P1 = 1
Upper Threshold	IUTH4P1 = ∞
Slope	ISL4P1 = −1.0

5. Spurious Stops by B

Score	IS5P1 = SSF(ILTH5P1,IB5P1,ISL5P1,IJS++)
Lower Threshold	ILTH5P1 = 0
Baseline	IB5P1 = 1
Upper Threshold	IUTH5P1 = ∞
Slope	ISL5P1 = −1.0

6. Availability

    Score   IS6P1 = SSF(ILTH6P1,IB6P1,IUTH6P1,
                                       ISL6P1,RLS[0,1])

    Lower Threshold   ILTH6P1 = 0

    Baseline   IB6P1 = 0.5

    Upper Threshold   IUTH6P1 = 1

    Slope   ISL6P1 = 2

RLS[0,1] indicates real numbers between 0 and 1, inclusive.

---

7. Reliability

    Score   IS7P1 = SSF(ILTH7P1,IB7P1,IUTH7P1,
                                       ISL7P1,RLS[0,1])

    Lower Threshold   ILTH7P1 = 0

    Baseline   IB7P1 = 0.5

    Upper Threshold   IUTH7P1 = 1

    Slope   ISL7P1 = 2

6.3.4.3   Weighting criteria

The following importance values, on a scale from 1 to 10, were assigned to each Performance Figure of Merit. These were provided by Dr. Bahill. The resultant weight, IWiP1, is computed by summing all the importance values and dividing each entry by this total. That is, the weights are normalized so that they add up to 1.0.

Figure of Merit	Value	IWiP1
1. Number of Collisions	8	0.258065
2. Trips by Train A	7	0.225806
3. Trips by Train B	7	0.225806
4. Spurious stops by A	3	0.096774
5. Spurious stops by B	3	0.096774
6. Availability	2	0.064516
7. Reliability	1	0.032258

6.3.5 *Utilization of Resources Requirement*

6.3.5.1 *Definition of Resource Figures of Merit*

The overall utilization of resources figure of merit is denoted UF0P1 and is computed by

$$UF0P1 = USF1P1 * UW1P1 + USF2P1 * UW2P1 + \ldots + USFnP1 * UWnP1$$

where n is the total number of Utilization of Resources Figures of Merit and

$$USFiP1 = USiP1(IFiP1(BSD)) \text{ for } i = 1 \text{ to } n$$

6.3.5.2 *Lower, upper, baseline, and scoring parameters*

In this section, the following naming convention for variables is used:

UFiP1 = the ith Utilization of Resources figure of merit,
UBiP1 = baseline value for the ith figure of merit,
ULTHiP1 = lower threshold for the ith figure of merit,
USFiP1 = score for the ith figure of merit,
USiP1 = scoring function for the ith figure of merit,
USLiP1 = slope for the ith figure of merit,
UUTHiP1 = upper threshold for the ith figure of merit,
UWiP1 = weight of the ith figure of merit, and
SSF = standard scoring function.

1. **Completion Time**

Score	US1P1 = SSF(ULTH1P1,UB1P1,USL1P1,IJS++)
Lower Threshold	ULTH1P1 = −1
Baseline	UB1P1 = 0
Upper Threshold	UUTH1P1 = ∞
Slope	USL1P1 = 1

2. **Acquisition Cost**

Score	US2P1 = SSF(ULTH2P1,UB2P1,UUTH2P1, USL2P1,RLS)
Lower Threshold	ULTH2P1 = 0
Baseline	UB2P1 = 0.5
Upper Threshold	UUTH2P1 = 1
Slope	USL2P1 = 2

2.1. Cost, Quantity 1

Score	US2.1P1 = SSF(ULTH2.1P1,UB2.1P1,USL2.1P1,RLS)
Lower Threshold	ULTH2.1P1 = 0
Baseline	UB2.1P1 = 250
Upper Threshold	UUTH2.1P1 = ∞
Slope	USL2.1P1 = −0.004

2.2. Cost, Quantity 1000

Score	US2.2P1 = SSF(ULTH2.2P1,UB2.2P1,USL2.2P1,RLS)
Lower Threshold	ULTH2.2P1 = 0
Baseline	UB2.2P1 = 100
Upper Threshold	UUTH2.2P1 = ∞
Slope	USL2.2P1 = −0.01

2.3. Cost, Quantity 1,000,000

Score	US2.3P1 = SSF(ULTH2.3P1,UB2.3P1,USL2.3P1,RLS)
Lower Threshold	ULTH2.3P1 = 0
Baseline	UB2.3P1 = 20
Upper Threshold	UUTH2.3P1 = ∞
Slope	USL2.3P1 = −0.05

6.3.5.3 *Weighting criteria*

The following importance values, on a scale from 1 to 10, were assigned by the customer to each Utilization of Resources Figure of Merit. The resultant weight, UW_iP1, is computed by summing all the importance values and dividing each entry by this total.

Figure of Merit	Value	UWiP1
1. Completion Time	10	0.5
2. Acquisition Cost	10	0.5
2.1. Cost, Quantity 1	3	0.333333
2.2. Cost, Quantity 1000	3	0.333333
2.3. Cost, Quantity 1,000,000	3	0.333333

6.3.6 *Trade-Off Requirement*

The Trade-Off Requirement is computed by the formula

$$TF0P1 = TW1P1 * IF0P1 + TW2P1 * UF0P1$$

where TW1P1 is the weight of the Overall I/O Performance Index and TW2P1 is the weight of the overall Utilization of Resources Index.

$$TW1P1 = 0.5$$
$$TW2P1 = 0.5$$

6.3.7 *System Test Requirement*

6.3.7.1 *Test plan*

6.3.7.1.1 *Explanation of test plan.*   The tests will be conducted by the TAs in the SIE-370 lab. The tests will be used to determine if the system requirements have been met.

Each system will be subjected to five separate tests, each having a two minute duration. The tests will run the given input trajectories with the given starting conditions, and the system must match the output trajectories.

For these tests, the term "danger zone" means the portions of the track between S1 and S2 and between S3 and S4.

6.3.7.1.2 *Test Trajectory 1.*   For Test Trajectory 1, Train B stops when Train A is within the danger zone, then Train B restarts when Train A leaves the danger zone. Both trains should start in the safe area of the track, but Train A should be positioned closer to the danger zone. The power to each train should be set to 75% of maximum.

TZ	IZ	OZ	Comment
0	(0,0,0,0)	(1,1)	/*both start in safe area*/
t(1)	(1,0,0,0)	(1,1)	/*S1 activated */
t(2)	(0,0,0,0)	(1,1)	/*Train A in danger zone */
t(3)	(0,0,1,0)	(1,1)	/*S3 activated */
t(4)	(0,0,1,0)	(1,0)	/*Train B stops */
t(5)	(0,1,1,0)	(1,0)	/*S2 activated */
t(6)	(0,0,1,0)	(1,1)	/*Train B starts */
t(7)	(0,0,0,0)	(1,1)	/*both trains run */
t(8)	(0,0,0,1)	(1,1)	/*S4 activated */
t(9)	(0,0,0,0)	(1,1)	/*both trains run */
•			
•			

Test continues for two minutes, allowing trains to generate their own trajectories.

The time t(i) is determined at the time of the test, the above time elements following the rule: t(i + 1) > t(i).

6.3.7.1.3 Test Trajectory 2. For Test Trajectory 2, Train A stops when Train B is within the danger zone, then Train A restarts when Train B leaves the danger zone. Both trains should start in the safe area of the track, but Train B should be positioned closer to the danger zone. The power to each train should be set to 75% of maximum.

TZ	IZ	OZ	Comment
0	(0,0,0,0)	(1,1)	/*both start in safe area*/
t(1)	(0,0,1,0)	(1,1)	/*S3 activated */
t(2)	(0,0,0,0)	(1,1)	/*Train B in danger zone */
t(3)	(1,0,0,0)	(1,1)	/*S1 activated */
t(4)	(1,0,0,0)	(0,1)	/*Train A stops */
t(5)	(1,0,0,1)	(0,1)	/*S4 activated */
t(6)	(1,0,0,0)	(1,1)	/*Train A starts */
t(7)	(0,0,0,0)	(1,1)	/*both trains run */
t(8)	(0,1,0,0)	(1,1)	/*S2 activated */
t(9)	(0,0,0,0)	(1,1)	/*both trains run */
.			
.			

Test continues for two minutes, allowing trains to generate their own trajectories.

The time $t(i)$ is determined at the time of the test, the above time elements following the rule: $t(i + 1) > t(i)$.

6.3.7.1.4 Test Trajectory 3. Test Trajectory 3 determines what happens when Test Trajectories 1 and 2 are run together. Both trains should start in the safe area of the track, but Train A should be positioned closer to the danger zone so that it will enter the danger zone before Train B. The power to Train A should be set to 25% of maximum and the power to Train B should be set to 75%.

TZ	IZ	OZ	Comment
0	(0,0,0,0)	(1,1)	/*both start in safe area */
t(1)	(1,0,0,0)	(1,1)	/*S1 activated */
t(2)	(0,0,0,0)	(1,1)	/*Train A in danger zone */
t(3)	(0,0,1,0)	(1,1)	/*S3 activated */
t(4)	(0,0,1,0)	(1,0)	/*Train B stops */
t(5)	(0,1,1,0)	(1,0)	/*S2 activated */
t(6)	(0,0,1,0)	(1,1)	/*Train B starts */
t(7)	(0,0,0,0)	(1,1)	/*both trains run */
t(8)	(0,0,0,1)	(1,1)	/*S4 activated */
t(9)	(0,0,0,0)	(1,1)	/*both trains run */
t(10)	(0,0,0,0)	(1,1)	/*both trains in safe area*/
t(11)	(0,0,1,0)	(1,1)	/*S3 activated */
t(12)	(0,0,0,0)	(1,1)	/*Train B in danger zone */
t(13)	(1,0,0,0)	(1,1)	/*S1 activated */
t(14)	(1,0,0,0)	(0,1)	/*Train A stops */
t(15)	(1,0,0,1)	(0,1)	/*S4 activated */
t(16)	(1,0,0,0)	(1,1)	/*Train A starts */
t(17)	(0,0,0,0)	(1,1)	/*both trains run */
t(18)	(0,1,0,0)	(1,1)	/*S2 activated */
t(19)	(0,0,0,0)	(1,1)	/*both trains run */

.

.

Test continues for two minutes, allowing trains to generate their own trajectories.

The time $t(i)$ is determined at the time of the test, the above time elements following the rule: $t(i + 1) > t(i)$.

6.3.7.1.5 Test Trajectory 4. Test Trajectory 4 determines what happens when Test Trajectories 2 and 1 are run together. Both trains should start in the safe area of the track, but this time Train B should be positioned closer to the danger zone so that it will enter the danger zone before Train A. The power to Train B should be set to 25% of maximum and the power to Train A should be set to 75%.

TZ	IZ	OZ	Comment
0	(0,0,0,0)	(1,1)	/*both start in safe area */
t(1)	(0,0,1,0)	(1,1)	/*S3 activated */
t(2)	(0,0,0,0)	(1,1)	/*Train B in danger zone */
t(3)	(1,0,0,0)	(1,1)	/*S1 activated */
t(4)	(1,0,0,0)	(0,1)	/*Train A stops */
t(5)	(1,0,0,1)	(0,1)	/*S4 activated */
t(6)	(1,0,0,0)	(1,1)	/*Train A starts */
t(7)	(0,0,0,0)	(1,1)	/*both trains run */
t(8)	(0,1,0,0)	(1,1)	/*S2 activated */
t(9)	(0,0,0,0)	(1,1)	/*both trains run */
t(10)	(0,0,0,0)	(1,1)	/*both trains in safe area*/
t(11)	(1,0,0,0)	(1,1)	/*S1 activated */
t(12)	(0,0,0,0)	(1,1)	/*Train A in danger zone */
t(13)	(0,0,1,0)	(1,1)	/*S3 activated */
t(14)	(0,0,1,0)	(1,0)	/*Train B stops */
t(15)	(0,1,1,0)	(1,0)	/*S2 activated */
t(16)	(0,0,1,0)	(1,1)	/*Train B starts */
t(17)	(0,0,0,0)	(1,1)	/*both trains run */
t(18)	(0,0,0,1)	(1,1)	/*S4 activated */
t(19)	(0,0,0,0)	(1,1)	/*both trains run */

Test continues for two minutes, allowing trains to generate their own trajectories.

The time $t(i)$ is determined at the time of the test, the above time elements following the rule: $t(i + 1) > t(i)$.

6.3.7.1.6 Test Trajectory 5. Test Trajectory 5 determines what happens when both trains enter the danger zone simultaneously. This test will have to be conducted on the functional model rather than on a physical model, with the results obtained through simulation.

TZ	IZ	OZ	Comment
0	(0,0,0,0)	(1,1)	`/*both start in safe area        */`
t(1)	(1,0,1,0)	(1,1)	`/*S1 & S3 activated               */`
t(2)	(0,0,1,0)	(1,0)	`/*Train B stop, Train A danger*/`
t(3)	(0,0,1,0)	(1,0)	`/*Train A in danger zone         */`
t(4)	(0,1,1,0)	(1,0)	`/*S2 activated                    */`
t(5)	(0,0,1,0)	(1,1)	`/*both trains go                  */`
t(6)	(0,0,0,0)	(1,1)	`/*Train B in danger zone         */`
t(7)	(0,0,0,1)	(1,1)	`/*S4 activated                    */`
t(8)	(0,0,0,0)	(1,1)	`/*both trains run                 */`
.			
.			

Test continues for two minutes, allowing trains to generate their own trajectories.

The time t(i) is determined at the time of the test, the above time elements following the rule: t(i + 1) > t(i).

6.3.7.2 Input/output performance tests

1. *Number of Collisions:* This figure of merit will be observed visually by the TA during the test period. Every train collision is noted. If the TA observes that a collision occurs because of irregular operation of the trains, the track, or the detectors, then the results will be voided and the test repeated.

2. *Trips by Train A:* This figure of merit will be observed visually by the TA during the test period. Every complete lap by Train A is noted. If the TA observes irregular operation of the trains, the track, or the detectors that results in fewer trips for Train A, then the results will be voided and the test repeated.

3. *Trips by Train B:* This figure of merit will be observed visually by the TA during the test period. Every complete lap by Train B is noted. If the TA observes irregular operation of the trains, the track, or the detectors that results in fewer trips for Train B, then the results will be voided and the test repeated.

4. *Spurious Stops by A:* This figure of merit will be observed visually by the TA during the test period. Every stop by Train A not needed to avoid a collision is noted. If the TA observes irregular operation of the trains, the track, or the detectors that results in a spurious stop by Train A, then the results will be voided and the test repeated.

5. *Spurious Stops by B:* The figure of merit will be observed visually by the TA during the test period. Every stop by Train B that was not needed to avoid a collision is noted. If the TA observes irregular operation of the trains, the track, or the detectors that results in a spurious stop by Train B, then the results will be voided and the test repeated.

6. *Availability:* The figure of merit will be observed visually by the TA at the beginning of the test. If the system works properly initially by giving obvious signs that it takes input from the detectors and sends outputs to the train controllers, then the system will be available and a figure of merit value of 1 is recorded. If not, a figure of merit value of 0 is recorded.

7. *Reliability:* The figure of merit will be observed by the TA throughout the entire length of the test. If the system works properly by giving obvious signs that it takes input from the detectors and sends output to the train controllers and if the system does not lose control of the outputs throughout the test, then the system will be reliable and a figure of merit value of 1 is recorded. If not, then a figure of merit value of 0 is recorded.

6.3.7.3   Utilization of resources tests

1. *Completion Time:* This figure of merit is observed by the instructor. The number of hours the project is completed before the due date is recorded as the figure of merit. The minimum value is –1, which indicates that the project was completed 1 hour late.

2. *Acquisition Cost:* This figure of merit is computed using the results of the next three figures of merit in the following equation

$$UF2P1(BSDi) = UW3P1 * US3P1(UF3P1(BSDi))$$
$$+ UW4P1 * US4P1(UF4P1(BSDi))$$
$$+ UW5P1 * US5P1(UF5P1(BSDi))$$

where BSDi is the buildable system design for concept i.

   2.1. *Cost, Quantity 1:* This figure of merit is an approximation of the cost of producing one of each product created in Document 7. These estimates are made by the student and, if judged reasonable by the instructor based on past estimates, they will be used as the figures of merit. If the student's figures of merit are not judged reasonable, the instructor will provide reasonable figures. If no estimates are submitted by the student, figure of merit values of 0 are recorded.

    2.2. *Cost, Quantity 1000:* This figure of merit is an approximation of the cost of producing 1000 of each product created in Document 7. These estimates are made by the student and, if judged reasonable by the instructor based on past estimates, they will be used as the figures of merit. If the student's figures of merit are not judged reasonable, the instructor will provide reasonable figures. If no estimates are submitted by the student, figure of merit values of 0 are recorded.

    2.3. *Cost, Quantity 1,000,000:* This figure of merit is an approximation of the cost of producing 1,000,000 of each product created in Document 7. These estimates are made by the student and, if judged reasonable by the instructor based on past estimates, they will be used as the figures of merit. If the student's figures of merit are not judged reasonable, the instructor will provide reasonable figures. If no estimates are submitted by the student, figure of merit values of 0 are recorded.

6.3.8   *Rationale for system requirements*

Data for the system requirements were provided by Dr. Bahill and the SIE-650 class.

6.4 Document 4: System Requirements Validation

In the System Requirements Validation Document we:

(1) examine the mathematical description of the Input/Output and Functional Requirements presented in Document 3 to check for consistency,

(2) demonstrate that a real-world solution can be built, and

(3) show that a real-world solution can be tested to prove that it satisfies the Input/Output and Functional Requirements.

If the client has requested a perpetual motion machine or a system that reduces entropy without expending energy, this is the place to stop the project and save the money.

6.4.1 Input/output and functional design

6.4.1.1 Terminology used

```
Z = (SZ, IZ, OZ, NZ, RZ)
```

where

Z = the system model of an I/O functional design,
SZ = states of the system,
IZ = inputs to the system,
OZ = outputs of the problem,
NZ = next state function, and
RZ = readout function.

Figure 6.3 shows a state diagram of this system.

6.4.1.2 States

There are three elements in the set of states:

1. Both trains are in a safe zone (SAFE),
2. Train A is in the intersection and Train B is in the safe zone (AinBout), and
3. Train B is in the intersection and Train A is outside it (BinAout).

```
SZ = {SAFE,AinBout,BinAout}
```

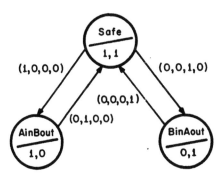

Figure 6.3 Diagram of a functional system design.

6.4.1.3 Inputs

There are four input ports to our system, which correspond to the four location detectors.

```
IZ = I1Z × I2Z × I3Z × I4Z
```

where

```
I1Z = {0,1} /*Input Port 1 represents Switch 1*/
            /*Train A not present = 0, present = 1*/

I2Z = {0,1} /*Input Port 2 represents Switch 2*/
            /*Train A not present = 0, present = 1*/

I3Z = {0,1} /*Input port 3 represents Switch 3*/
            /*Train B not present = 0, present = 1*/

I4Z = {0,1} /*Input port 4 represents Switch 4*/
            /*Train B not present = 0, present = 1*/
```

6.4.1.4 Outputs

There are two output ports, which correspond to the power connections to the trains.

```
OZ = O1Z × O2Z
```

where

```
O1Z = {0,1} /*Output port 1, O1Z, represents
             power to Train A.
             Power to A OFF = 0, ON = 1*/

O2Z = {0,1} /*Output port 2, O2Z, represents
             power to Train B.
             Power to B OFF = 0, ON = 1*/
```

6.4.1.5 *Next state function*

The next state function specifies the next state of the system given the present state of the system and the present inputs. It is arranged as {((state1, input),nextstate), ((state2,input),nextstate), ...} for every possible combination of starting state and inputs. For the train controller we have:

```
NZ = {(SAFE,(1,0,0,0),AinBout),
      (SAFE,(0,0,1,0),BinAout),
      (SAFE,(1,0,1,0),AinBout),
      (AinBout,(0,1,0,0),SAFE),
      (BinAout,(0,0,0,1),SAFE)}
   U {((SAFE,p),SAFE): p ∈ IZ;
        p <> {(1,0,0,0),(0,0,1,0),(1,0,1,0)}}
   U {((AinBout,p),AinBout): p ∈ IZ;
        p <> {(0,1,0,0)}}
   U {((BinAout,p),BinAout): p ∈ IZ;
        p <> {(0,0,0,1)}}
```

6.4.1.6 *Readout function*

The readout function specifies the values of the outputs for each state. Its form is `RZ = {(state1,output1), (state2,output2), ...}`. For the train controller this becomes:

```
RZ = {(SAFE,(1,1)),(AinBout,(1,0)),
      (BinAout,(0,1))}
```

The requirements state that power must be ON to one of the trains at all times; therefore, output (0,0) is not allowed.

6.4.2 *A feasible system design*

A software system can be designed to emulate the functional design. This software system would then interface with the train controllers via the Motorola controller mentioned in the System Requirements Document.

6.4.3 *A real system*

This system has been implemented by 300 students in the last five years, with many of their solutions meeting the acceptance criteria. Therefore, it is reasonable to expect that the system can be implemented once again.

6.5 *Document 5: Concept Exploration*

> The Concept Exploration Document is used to study several different system designs via approximation, simulation, or experiments with prototypes. The best design alternative is suggested by the data. This document will be rewritten many times as more information becomes available.

6.5.1 *System design concepts*

6.5.1.1 *System Design Concept 1*

6.5.1.1.1 Explanation of System Design Concept 1.   Concept 1 specifies the use of a hardware protoboard to control the trains, as described in systems engineering Document 3, System Requirements. The board will be built with eight pins on its edge for the inputs of the train location and the outputs for power. They will be in the following order: S1, S2, S3, S4, GND, GND, PA, and PB, with S1 on pin 1 of the connector, S2 on pin 2, and so on.

The functions will be implemented using TTL integrated circuits and other components available in the SIE-370 lab. Boards will be handwired and then delivered to the TA in the lab for testing.

6.5.1.1.2 Model of System Design Concept 1

6.5.1.1.2.1 Terminology used

$$Z1 = (SZ1, IZ1, OZ1, NZ1, RZ1)$$

where

$Z1$ = the model of I/O Functional Design Concept 1,
$SZ1$ = states of system $Z1$,
$IZ1$ = inputs to system $Z1$,
$OZ1$ = outputs of system $Z1$,
$NZ1$ = next state function, and
$RZ1$ = readout function.

6.5.1.1.2.2 States

$$SZ1 = \{SAFE, AinBout, BinAout\}$$

6.5.1.1.2.3 Inputs

```
IZ1 = I1Z1 × I2Z1 × I3Z1 × I4Z1
   I1Z1 = {0,1} /*where 1 indicates S1 activated*/
   I2Z1 = {0,1} /*where 1 indicates S2 activated*/
```

```
I3Z1 = {0,1} /*where 1 indicates S3 activated*/
I4Z1 = {0,1} /*where 1 indicates S4 activated*/
```

6.5.1.1.2.4 Outputs

```
OZ1 = O1Z1 × O2Z1 ·
   O1Z1 = {0,1} /*where 1 indicates power ON to
                   train A*/
   O2Z1 = {0,1} /*where 1 indicates power ON to
                   train B*/
```

6.5.1.1.2.5 Next state function

```
NZ1 = {(SAFE,(1,0,0,0),AinBout),
       (SAFE,(0,0,1,0),BinAout),
       (SAFE,(1,0,1,0),AinBout),
       (AinBout,(0,1,0,0),SAFE),
       (BinAout,(0,0,0,1),SAFE)}
    ∪ {((SAFE,p),SAFE): p ∈ IZ;
        p <> {(1,0,0,0),(0,0,1,0),(1,0,1,0)}}
    ∪ {((AinBout,p),AinBout): p ∈ IZ;
        p <> {(0,1,0,0)}}
    ∪ {((BinAout,p),BinAout): p ∈ IZ;
        p <> {(0,0,0,1)}}
```

In this function, p represents the present value of the inputs. The input $(0,0,1,0)$ indicates that Switch 1 is 0, Switch 2 is 0, Switch 3 is 1, and Switch 4 is 0. If p does not equal this or the other listed combinations, the system remains in the state S A F E. In this case, we listed in detail the first combinations, then we grouped together other combinations and joined them to the prior set with a union.

6.5.1.1.2.6 Readout function

```
RZ1 = {(SAFE,(1,1)),(AinBout,(1,0)),
       (BinAout,(0,1))}
```

This is a simple readout function that says: When in state B i n A o u t, set Output Port 1 to 0 (Train A OFF) and Output Port 2 to 1 (Train B ON). We leave it to the student to ascertain whether or not we satisfied the matching function with this system.

6.5.1.2 System Design Concept 2

6.5.1.2.1 Explanation of System Design Concept 2. System Design Concept 2 specifies the use of an assembly language program to control the trains. The program will interface to the outside world via a parallel port on the back of the Motorola controller in the lab. The program can be written on one of the Unix-based PCs or DOS-based IBM PCs in the lab. After being properly debugged, the program will be downloaded to the Motorola controller.

6.5.1.2.2 ˙ Model of System Design Concept 2

6.5.1.2.2.1 Terminology used

$$Z2 = (SZ2, IZ2, OZ2, NZ2, RZ2)$$

where

$Z2$ = the model of I/O Functional Design Concept 2,
$SZ2$ = states of system $Z2$,
$IZ2$ = inputs to system $Z2$,
$OZ2$ = outputs of system $Z2$,
$NZ2$ = next state function, and
$RZ2$ = readout function.

Concept 2 is functionally identical to Concept 1, the only difference being the variable name—Concept 2 is referred to as $Z2$ instead of $Z1$.

6.5.1.3 System Design Concept 3

6.5.1.3.1 Explanation of System Design Concept 3 Concept 3 specifies the use of a computer program written in a high-level computer language to control the trains. The program will interface to the outside world via a parallel port on the back of the Motorola controller in the lab. The code may be written in Pascal on one of the personal computers in the lab. After the code is compiled (the final step in writing a program), an assembler is used to create an executable version that can be downloaded to the Motorola controller.

6.5.1.3.2 Model of System Design Concept 3

6.5.1.3.2.1 Terminology used

$$Z3 = (SZ3, IZ3, OZ3, NZ3, RZ3)$$

where

$Z3$ = the model of I/O Functional Design Concept 3,
$SZ3$ = states of system $Z3$,
$IZ3$ = inputs to system $Z3$,
$OZ3$ = outputs of system $Z3$,
$NZ3$ = next state function, and
$RZ3$ = readout function.

Concept 3 is functionally identical to Concept 1, the only difference being the variable name—Concept 3 is referred to as **Z3** instead of **Z1**.

6.5.2   Figures of merit

The figures of merit are calculated using the test plan described in Document 3. The values obtained for these figures of merit are entered here, then the scores are computed using the standard scoring functions also defined in Document 3. The formulas

$$IF0P1(FSDi) = IW1P1 * ISF1P1 + \dots + IWmP1 * ISFmP1$$
$$UF0P1(FSDi) = UW1P1 * USF1P1 + \dots + UWnP1 * USFnP1$$

are used to compute the Overall Figures of Merit for each design, where m is the number of Input/Output Performance Figures of Merit and n is the number of Utilization of Resources Figures of Merit; and

$$ISF1P1 = IS1P1(IF1P1(FSDi))$$
$$USF1P1 = US1P1(UF1P1(FSDi))$$

where i is the concept design number. IF1P1(FSDi) is the measured Performance Figures of Merit for Requirement 1 for Functional System Design Concept i. This is entered into the scoring function IS1P1 to generate the scaled score ISF1P1.

The tables on the following pages show the estimates given for the figures of merit based on the models developed in Documents 6 and 7. The column titled IFiP1 (where i is the figure of merit number) is the figure of merit measured per the test plan for Functional System Design 1. The column labeled ISFiP1 is the calculated score after entering the figure of merit into the standard scoring function defined in Document 3. The column IWiP1 is the weight factor given in Document 3 for the respective figure of merit. The overall scores, IF0P1 and UF0P1, are determined from the weights and scores.

Three different methods for determining the figures of merit are given: approximation, simulation, and prototype. These methods reflect the different types of data available for determining figures of merit throughout the initial design. Approximation values are "blue sky" guesses by Dr. Bahill based on his experience and on historical data. Simulation data are obtained using models built to simulate the prototype. Prototype data are actual measurements on a working prototype.

Concept 1 specifies the use of a hardware protoboard to control the trains. Tables for the approximation, simulation, and prototype methods follow.

6.5.2.1.1 Approximation figures of merit for Concept 1

I/O FIGURES OF MERIT

REQUIREMENTS	IFiP1(FSD1)	ISFiP1(FSD1)	IWiP1
1 Number of Collisions	0	1	0.258065
2 Trips by Train A	7	0.832	0.225806
3 Trips by Train B	5	0.5	0.225806
4 Spurious Stops by A	0	1	0.096774
5 Spurious Stops by B	1	0.5	0.096774
6 Availability	1	1	0.064516
7 Reliability	1	1	0.032258

IF0P1(FSD1) = 0.801

U/R FIGURES OF MERIT

REQUIREMENTS	UFiP1(FSD1)	USFiP1(FSD1)	UWiP1
1 Completion Time	2	1	0.5
2 Acquisition Cost	0.895	0.978	0.5
2.1 Cost, Quantity 1	44	0.983	0.333333
2.2 Cost, Quantity 1000	25	0.97	0.333333
2.3 Cost, Quantity 1,000,000	15	0.732	0.333333

UF0P1(FSD1) = 0.989

6.5.2.1.2 Simulation figures of merit for Concept 1

I/O FIGURES OF MERIT

REQUIREMENTS	IFiP1(FSD1)	ISFiP1(FSD1)	IWiP1
1 Number of Collisions	0	1	0.258065
2 Trips by Train A	8	0.917	0.225806
3 Trips by Train B	8	0.917	0.225806
4 Spurious Stops by A	0	1	0.096774
5 Spurious Stops by B	0	1	0.096774
6 Availability	1	1	0.064516
7 Reliability	1	1	0.032258

IF0P1(FSD1) = 0.963

U/R FIGURES OF MERIT

REQUIREMENTS	UFiP1(FSD1)	USFiP1(FSD1)	UWiP1
1 Completion Time	0	0.5	0.5
2 Acquisition Cost	0.895	0.978	0.5
2.1 Cost, Quantity 1	44	0.983	0.333333
2.2 Cost, Quantity 1000	25	0.97	0.333333
2.3 Cost, Quantity 1,000,000	15	0.732	0.333333

UF0P1(FSD1) = 0.739

6.5.2.1.3 Prototype figures of merit for Concept 1

I/O FIGURES OF MERIT

REQUIREMENTS	IFiP1(FSD1)	ISFiP1(FSD1)	IWiP1
1 Number of Collisions	0	1	0.258065
2 Trips by Train A	8	0.917	0.225806
3 Trips by Train B	8	0.917	0.225806
4 Spurious Stops by A	1	0.5	0.096774
5 Spurious Stops by B	1	1	0.096774
6 Availability	1	1	0.064516
7 Reliability	1	1	0.032258

IF0P1(FSD1) = 0.914

U/R FIGURES OF MERIT

REQUIREMENTS	UFiP1(FSD1)	USFiP1(FSD1)	UWiP1
1 Completion Time	0	0.5	0.5
2 Acquisition Cost	0.895	0.978	0.5
2.1 Cost, Quantity 1	44	0.983	0.333333
2.2 Cost, Quantity 1000	25	0.97	0.333333
2.3 Cost, Quantity 1,000,000	15	0.732	0.333333

UF0P1(FSD1) = 0.739

6.5.2.2 *Figures of merit for Concept 2*

Concept 2 specifies the use of an assembly language program to control the trains. Tables for the approximation, simulation, and prototype methods follow.

6.5.2.2.1 Approximation figures of merit for Concept 2

I/O FIGURES OF MERIT

REQUIREMENTS	IFiP1(FSD2)	ISFiP1(FSD2)	IWiP1
1 Number of Collisions	0	1	0.258065
2 Trips by Train A	8	0.917	0.225806
3 Trips by Train B	8	0.917	0.225806
4 Spurious Stops by A	0	1	0.096774
5 Spurious Stops by B	0	1	0.096774
6 Availability	1	1	0.064516
7 Reliability	1	1	0.032258

IF0P1(FSD2) = 0.963

U/R FIGURES OF MERIT

REQUIREMENTS	UFiP1(FSD2)	USFiP1(FSD2)	UWiP1
1 Completion Time	5	1	0.5
2 Acquisition Cost	0.013	0.001	0.5
2.1 Cost, Quantity 1	450	0.039	0.333333
2.2 Cost, Quantity 1000	350	0	0.333333
2.3 Cost, Quantity 1,000,000	200	0	0.333333

UF0P1(FSD2) = 0.5

6.5.2.2.2 Simulation figures of merit for Concept 2

I/O FIGURES OF MERIT

REQUIREMENTS	IF*i*P1(FSD2)	ISF*i*P1(FSD2)	IW*i*P1
1 Number of Collisions	0	1	0.258065
2 Trips by Train A	8	0.917	0.225806
3 Trips by Train B	8	0.917	0.225806
4 Spurious Stops by A	0	1	0.096774
5 Spurious Stops by B	0	1	0.096774
6 Availability	1	1	0.064516
7 Reliability	1	1	0.032258

IF0P1(FSD2) = 0.963

U/R FIGURES OF MERIT

REQUIREMENTS	UF*i*P1(FSD2)	USF*i*P1(FSD2)	UW*i*P1
1 Completion Time	0	0.5	0.5
2 Acquisition Cost	0.013	0.001	0.5
2.1 Cost, Quantity 1	450	0.039	0.333333
2.2 Cost, Quantity 1000	350	0	0.333333
2.3 Cost, Quantity 1,000,000	200	0	0.333333

UF0P1(FSD2) = 0.251

6.5.2.2.3   Prototype figures of merit for Concept 2

I/O FIGURES OF MERIT

REQUIREMENTS	IFiP1(FSD2)	ISFiP1(FSD2)	IWiP1
1 Number of Collisions	0	1	0.258065
2 Trips by Train A	9	0.961	0.225806
3 Trips by Train B	10	0.982	0.225806
4 Spurious Stops by A	0	1	0.096774
5 Spurious Stops by B	0	1	0.096774
6 Availability	1	1	0.064516
7 Reliability	1	1	0.032258

IF0P1(FSD2) = 0.987

U/R FIGURES OF MERIT

REQUIREMENTS	UFiP1(FSD2)	USFiP1(FSD2)	UWiP1
1 Completion Time	5	1	0.5
2 Acquisition Cost	0.013	0.001	0.5
2.1 Cost, Quantity 1	450	0.039	0.333333
2.2 Cost, Quantity 1000	350	0	0.333333
2.3 Cost, Quantity 1,000,000	200	0	0.333333

UF0P1(FSD2) = 0.5

6.5.2.3 Figures of Merit for Concept 3

Concept 3 specifies the use of a high-level language program to control the trains. Tables for the approximation and simulation methods follow.

6.5.2.3.1 Approximation figures of merit for Concept 3

I/O FIGURES OF MERIT

REQUIREMENTS	IF*i*P1(FSD3)	ISF*i*P1(FSD3)	IW*i*P1
1 Number of Collisions	0	1	0.258065
2 Trips by Train A	8	0.917	0.225806
3 Trips by Train B	8	0.917	0.225806
4 Spurious Stops by A	0	1	0.096774
5 Spurious Stops by B	0	1	0.096774
6 Availability	1	1	0.064516
7 Reliability	1	1	0.032258

IF0P1(FSD3) = 0.963

U/R FIGURES OF MERIT

REQUIREMENTS	UF*i*P1(FSD3)	USF*i*P1(FSD3)	UW*i*P1
1 Completion Time	5	1	0.5
2 Acquisition Cost	0.013	0.001	0.5
2.1 Cost, Quantity 1	450	0.039	0.333333
2.2 Cost, Quantity 1000	350	0	0.333333
2.3 Cost, Quantity 1,000,000	200	0	0.333333

UF0P1(FSD3) = 0.5

6.5.2.3.2   Simulation figures of merit for Concept 3

I/O FIGURES OF MERIT

REQUIREMENTS	IF*i*P1(FSD3)	ISF*i*P1(FSD3)	IW*i*P1
1 Number of Collisions	0	1	0.258065
2 Trips by Train A	8	0.917	0.225806
3 Trips by Train B	8	0.917	0.225806
4 Spurious Stops by A	0	1	0.096774
5 Spurious Stops by B	0	1	0.096774
6 Availability	1	1	0.064516
7 Reliability	1	1	0.032258

IF0P1(FSD3) = 0.963

U/R FIGURES OF MERIT

REQUIREMENTS	UF*i*P1(FSD3)	USF*i*P1(FSD3)	UW*i*P1
1 Completion Time	0	0.5	0.5
2 Acquisition Cost	0.013	0.001	0.5
2.1 Cost, Quantity 1	450	0.039	0.333333
2.2 Cost, Quantity 1000	350	0	0.333333
2.3 Cost, Quantity 1,000,000	200	0	0.333333

UF0P1(FSD3) = 0.251

6.5.3   Trade-off analysis

The trade-off analysis compares the different design concepts. After the figures of merit are collected and the scores computed, the Overall Perform-ance Figure of Merit and the Overall Utilization of Resources Figure of Merit are used to compute the trade-off scores for each category of figures of merit. Comparisons are made for the approximation, simulation, and prototype data. The symbology IF0P1(FSD1) indicates this is the Overall Input/Output Performance Figure of Merit for Problem 1 of SIERRA for the Functional System Design Concept 1.

6.5.3.1   Approximation trade-off analysis
The scores for the Input/Output Performance Requirement and the Utiliza-tion of Resources Requirement for the approximation data are summarized here with the Trade-Off Requirement.

Concept 1: Hardware prototype

$$TW1P1 * IF0P1(FSD1) + TW2P1 * UF0P1(FSD1) = TF0P1(FSD1)$$
$$0.5 \quad * \quad 0.801 \quad + \quad 0.5 \quad * \quad 0.989 \quad = \quad 0.895$$

Concept 2: Assembly language software

$$TW1P1 * IF0P1(FSD2) + TW2P1 * UF0P1(FSD2) = TF0P1(FSD2)$$
$$0.5 \quad * \quad 0.963 \quad + \quad 0.5 \quad * \quad 0.5 \quad = \quad 0.7315$$

Concept 3: High-level language software

$$TW1P1 * IF0P1(FSD3) + TW2P1 * UF0P1(FSD3) = TF0P1(FSD3)$$
$$0.5 \quad * \quad 0.963 \quad + \quad 0.5 \quad * \quad 0.5 \quad = \quad 0.7315$$

6.5.3.2 Simulation trade-off analysis
The scores for the Input/Output Performance Requirement and the Utilization of Resources Requirement for the simulation data are summarized here with the Trade-Off Requirement.

Concept 1: Hardware prototype

$$TW1P1 * IF0P1(FSD1) + TW2P1 * UF0P1(FSD1) = TF0P1(FSD1)$$
$$0.5 \quad * \quad 0.963 \quad + \quad 0.5 \quad * \quad 0.739 \quad = \quad 0.851$$

Concept 2: Assembly language software

$$TW1P1 * IF0P1(FSD2) + TW2P1 * UF0P1(FSD2) = TF0P1(FSD2)$$
$$0.5 \quad * \quad 0.963 \quad + \quad 0.5 \quad * \quad 0.251 \quad = \quad 0.607$$

Concept 3: High-level language software

$$TW1P1 * IF0P1(FSD3) + TW2P1 * UF0P1(FSD3) = TF0P1(FSD3)$$
$$0.5 \quad * \quad 0.963 \quad + \quad 0.5 \quad * \quad 0.251 \quad = \quad 0.607$$

6.5.3.3 Prototype trade-off analysis
The scores for the Input/Output Performance Requirement and the Utilization of Resources Requirement for the prototype data are summarized here with the Trade-Off Requirement.

Concept 1: Hardware prototype

$$TW1P1 * IF0P1(FSD1) + TW2P1 * UF0P1(FSD1) = TF0P1(FSD1)$$
$$0.5 \quad * \quad 0.914 \quad + \quad 0.5 \quad * \quad 0.739 \quad = \quad 0.8265$$

Concept 2:   Assembly language software

$$TW1P1 * IF0P1(FSD2) + TW2P1 * UF0P1(FSD2) = TF0P1(FSD2)$$
$$0.5 \quad * \quad 0.987 \quad + \quad 0.5 \quad * \quad 0.5 \quad = \quad 0.7435$$

6.5.4   Alternative recommended

6.5.4.1   Approximation alternatives
Concept 1 had an overall score of 0.895, Concept 2 had a score of 0.7315, and Concept 3 had a score of 0.7315. Thus, by the approximations of the system designers, the hardware protoboard is the best concept. Though the other systems were expected to function better, the hardware prototype system was chosen because of its lower cost.

6.5.4.2   Simulation alternatives
The simulations were done using computer software to model the functions of the system per the state diagrams given in Document 6. The scores were 0.851 for Concept 1, 0.607 for Concept 2, and 0.0607 for Concept 3. Again, the hardware protoboard came out ahead because of low cost.

6.5.4.3   Prototype alternatives
A prototype of Concept 3 could not be done because no compiler was available. The scores were 0.8265 for Concept 1 and 0.7435 for Concept 2. The hardware protoboard prototype experienced problems during the tests, which caused several spurious stops. The good results of the software, which allowed more trips for the trains, still did not outweigh their higher cost.

The hardware version was the best in all the analyses because of the low cost of resources. The software unit had to include the expense of the Motorola ECB, which seriously degraded its score.

> *If our boss was not satisfied with this answer and demanded an assembly language solution, we could change the weight of acquisition cost (IW2P1) from 0.5 to 0.05. This would cause the overall score of Concept 2 to be the highest. This kind of manipulation is often done in the real world—we are not just being cynical here. However, now there is documentation on the person responsible for the decision, and the system design can be traced to this decision. It is important to remember that the customer's weights are what is important; changing them would result in a non-optimal design from the customer's perspective.*

6.5.5 Sensitivity analysis

The I/O Performance Requirements scores were similar for all three concepts, with the software-controlled systems having a slight advantage. However, because a computer must be included with each software unit sold, the cost of the software system was too high. The performance results show that the software-controlled systems were more effective at completing laps than the hardware system, but not enough to make up for the difference in cost. Even a significant decrease in the cost of the Motorola ECB would not allow the software systems to compete with the hardware system.

6.5.6 Rationale for alternatives, models, and methods

There is little difference in the function of the models. Though the technology used in each is different, their functional performance in controlling the trains is almost identical, with the software systems being slightly better.

If the Acquisition Cost was given a considerably lower priority when the weights were determined in Document 3, the recommended system would have been different. Upon a reevaluation of the weights, no changes are made to them. They are considered valid, since the cost of the system is a significant consideration of any potential consumer. Therefore, the recommended alternative remains the hardware system.

The cost of one manufactured piece of the hardware was estimated as:

Protoboard	$30.00
Power Supply	5.00
ICs	5.00
Wire	2.00
LEDs	1.00
Supplies	1.00
Total	$44.00

The cost of the creation of one copy of the software was estimated as:

ECB	$450.00
Total	$450.00

Estimates for quantities of 1000 and 1,000,000 were based on volume purchases.

Other figures of merit that could be created which might change the recommended alternatives are:

(1) ease of troubleshooting,

(2) ease of modification,

(3) physical robustness, and

(4) appearance.

The customer may not think of all these criteria, so a part of the systems engineer's job is careful questioning of the customer to determine what is important in evaluating alternatives. In Chapter 7 we show how quality function deployment (QFD) can be used to help elicit all the customer's important criteria.

6.6 Document 6: System Functional Analysis

The System Functional Analysis Document decomposes the I/O and Functional Requirements into a functional system design. Its intended audience is systems engineers.

6.6.1 System functional analysis of Concept 1

6.6.1.1 Top level system functional analysis of Concept 1

Concept 1 uses hardware to implement the requirements. Based on Document 3 requirements and the Concept 1 states defined in Document 5, the major states of the system are:

1. SAFE: Both trains are in the safe zone. Power will be ON to both trains.

2. AinBout: Train A is in the danger zone; Train B is outside the danger zone. Power will be ON to Train A and OFF to Train B.

3. BinAout: Train B is in the danger zone; Train A is outside the danger zone. Power will be ON to Train B and OFF to Train A.

The system will start in SAFE. See Figure 6.4 for the top level state diagram. These states suggest the following functions:

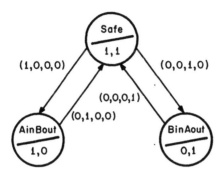

Figure 6.4 State diagram for Alternative-1.

1. Monitor sensors to see when either train enters the danger zone. We will call this Monitor13 to represent the function of monitoring Switches 1 and 3. This is from the SAFE state.

2. Monitor sensors to see if Train A leaves the danger zone or if Train B attempts to enter the danger zone. We will call this Monitor23. This is from AinBout.

3. Monitor sensors to see if Train B leaves the danger zone or if Train A attempts to enter the danger zone. We will call this Monitor14. This is from BinAout.

> *This system uses the initial states as a guideline for defining the functions. The functions are performed by the next state function and the readout function, but in this case the combination is unique to each state. Therefore, it is possible to pattern the functional decomposition after our states.*

6.6.1.2 Functional decomposition

6.6.1.2.1 Function 1. The first function of the system is Monitor13. Based on Document 3, the following initial conditions must be set:

1. Location outside of danger zone. This means initial input is
   IRP1 = (0,0,0,0).

2. Power ON to both trains. This means that initial output is
   ORP1 = (1,1).

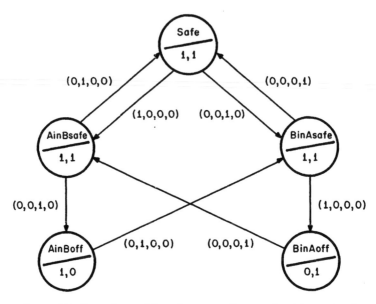

Figure 6.5 Complete subfunction state diagram for Alternative-1.

Therefore, the output from this function is power to both trains. This function is left if either Switch 1 is triggered (I1P1 = 1) or Switch 3 is triggered (I3P1 = 1). No further decomposition of this function is necessary.

   6.6.1.2.2 Function 2.   In the second function, Monitor23, Train A is in the danger zone and Train B is in the safe zone. This function is entered when Train A triggers S1 (I1P1 = 1) while in the Monitor13 function.

   To prevent collisions, this function will turn the power OFF to Train B. To increase the I/O Performance Requirement for the number of laps Train B completes, it is desirable to decompose this function into two subfunctions that allow Train B to have power until it attempts to enter the danger zone. The function Monitor23 will be decomposed into Monitor23On and Monitor23Off. These are patterned after states AinBsafe and AinBoff. See Figure 6.5 for the new state diagram. The input from Switch S1 starts function Monitor23On. Power is maintained to both trains. This function is exited and the Monitor23Off function is entered if Train B triggers S3 [input (0,0,1,0) is detected]. This will turn the power OFF to Train B. If S2 is triggered [input (0,1,0,0) is detected], the Monitor13 function resumes.

   6.6.1.2.3 Function 3.   In the third function, Monitor14, Train B is in the danger zone and Train A is in the safe zone. This function is entered when Train B triggers Switch 3 (I3P1 = 1), indicating that it wants to enter the danger zone also, while in the Monitor13 function.

   To prevent collisions, power to Train A is turned OFF. To increase the I/O Performance Requirement for the number of laps Train A completes, it is desirable to decompose this function into two subfunctions that allow Train A to have power until it attempts to enter the danger zone. The function is decomposed into Monitor14On and Monitor14Off. These are patterned after states BinAsafe and BinAoff. See Figure 6.5 for the new state diagram. An input to Monitor13 of (0,0,1,0) drives the system into the new function, Monitor14On. Power is maintained to both trains. This function is exited and Monitor14Off is entered if Train A triggers S1 [input (1,0,0,0) is detected]. Monitor14On goes to function Monitor13 if S4 is triggered [input (0,0,0,1) is detected].

6.6.1.3 Complete functional model

6.6.1.3.1 Terminology

```
Z1' = (SZ1', IZ1', OZ1', NZ1', RZ1')
```

where

   $Z1'$ = the model of a complete I/O functional design,
   $SZ1'$ = states of system $Z1'$,
   $IZ1'$ = inputs to system $Z1'$,
   $OZ1'$ = outputs of system $Z1'$,
   $NZ1'$ = next state function, and
   $RZ1'$ = readout function.

6.6.1.3.2 States

```
SZ1' = {SAFE,AinBsafe,AinBoff,BinAsafe,BinAoff}
```

6.6.1.3.3 Inputs

```
IZ1' = I1Z1 × I2Z1 × I3Z1 × I4Z1
   I1Z1 = {0,1} /*where 1 indicates S1 activated*/
   I2Z1 = {0,1} /*where 1 indicates S2 activated*/
   I3Z1 = {0,1} /*where 1 indicates S3 activated*/
   I4Z1 = {0,1} /*where 1 indicates S4 activated*/
```

6.6.1.3.4 Outputs

```
OZ1' = O1Z1 × O2Z1
   O1Z1 = {0,1} /*where 1 indicates power ON to
                  Train A*/
   O2Z1 = {0,1} /*where 1 indicates power ON to
                  Train B*/
```

6.6.1.3.5 Next state function

```
NZ1' = {(SAFE,(1,0,0,0),AinBsafe),
        (SAFE,(0,0,1,0),BinAsafe),
        (AinBsafe,(0,1,0,0),SAFE),
        (AinBsafe,(0,0,1,0),AinBoff),
```

```
(AinBoff,(0,1,0,0),BinAsafe)
(BinAsafe,(0,0,0,1),SAFE),
(BinAsafe,(1,0,0,0),BinAoff),
(BinAoff,(0,0,0,1),AinBsafe)}
```

All unlisted input combinations cause the system to remain in the same state.

6.6.1.3.6   Readout Function

```
RZ1' = {(SAFE,(1,1)),(AinBsafe,(1,1)),
        (AinBoff,(1,0)),(BinAsafe,(1,1)),
        (BinAoff,(0,1))}
```

6.6.1.4   Rationale for analysis of Concept 1

Concept 1 is based on the requirements from Document 3 and the Concept 1 portion of Document 5. A computer simulation was run to test the complete functional decomposition using the items described in Document 3. Exhibit 6.1 is the simulation program listing, and Exhibit 6.2 is the result. The simulation verified the correctness of the complete state diagram. The model worked well and all the tests were passed correctly.

EXHIBIT 6.1

Simulation Program Listing

```
/*PSEUDO CODE FOR SIMULATION OF THE HARDWARE
                FUNCTIONAL DESIGN*/
/*----------------------------------------*/

/*Beginning of loop for test trajectory*/
for each test trajectory do
    put power on to each train per test
        trajectory
    position each train per test trajectory

/*Beginning of loop to get inputs and process
    them*/
    get next input of test trajectory until
            done
        if the current state is SAFE
        /*SAFE*/
            if the input is (x,x,x,1) then
            /* S4 */
                the next state is BinAsafe
                B is in danger
```

```
      else if the input is (1,x,x,x) then
      /* S1 */
         the next state is AinBsafe
         A is in danger
      else
      /*other*/
         the next state is SAFE
         B is safe
         A is safe         .
   else if the current state is AinBsafe
   /*AinBsafe*/
      if the input is (x,x,1,x) then
      /* S3 */
         the next state is AinBoff
            B has power off
            A is in danger
      else if the input is (x,1,x,x) then
      /* S2 */
         the next state is SAFE
         B is safe with power on
         A is safe with power on
      else
      /*other*/
         the next state is AinBsafe
         A is in danger
   else if the current state is BinAsafe
   /*BinAsafe*/
      if the input is (1,x,x,x) then
      /* S1 */
         the next state is BinAoff
            A has power off
            B is in danger
      else if the input is (x,x,x,1) then
      /* S4 */
         the next state is SAFE
         B is safe with power on
         A is safe with power on
      else
      /*other*/
         the next state is BinAsafe
         B is in danger
   else if the current state is AinBoff
   /*AinBoff*/
      if the input is (x,1,x,x) then
      /* S2 */
```

```
                    the next state is BinAsafe
                    B has power on
                    B is in danger
                    A is safe
              else
              /*other*/
                  the next state is AinBoff
                  B has power off
                  A is in danger
          else if the current state is BinAoff
          /*BinAoff*/
              if the input is (x,x,x,1) then
              /* S4 */
                  the next state is AinBsafe
                      A has power on
                      A is in danger
                      B is safe
              else
              /*other*/
                  the next state is BinAoff
                  A has power off
                  B is in danger
          end of loop for state
      if A is in danger and B is in danger then
      /*OUTPUT ROUTINE*/
          print COLLISION
      else if A is in danger then
          print A IN DANGER
          print B IS SAFE
      else if B is in danger then
          print B IN DANGER
          print A IS SAFE
      else
          print A IS SAFE
          print B IS SAFE
          end of output routine if statement
      end loop for test trajectory input
end loop for test trajectory
```

EXHIBIT 6.2

Simulation Results for Concept 1

These tables represent the output of the simulation code in Exhibit 6.1. The test trajectories from the Test Requirement were entered into the functional design to see how the system would perform.

Begin test of hardware conceptual design simulation.

Test Trajectory 1

TZ	State	Input	Output	Train A	Train B
0	SAFE	(0,0,0,0)	(1,1)	safe	safe
1	SAFE	(1,0,0,0)	(1,1)	S1	safe
2	AinBsafe	(0,0,0,0)	(1,1)	danger	safe
3	AinBsafe	(0,0,1,0)	(1,1)	danger	S3
4	AinBoff	(0,0,1,0)	(1,0)	danger	S3
5	AinBoff	(0,1,1,0)	(1,0)	S2	S3
6	BinAsafe	(0,0,1,0)	(1,1)	safe	S3
7	BinAsafe	(0,0,0,0)	(1,1)	safe	danger
8	BinAsafe	(0,0,0,1)	(1,1)	safe	S4
9	SAFE	(0,0,0,0)	(1,1)	safe	safe
10	SAFE				

Test Trajectory 2

TZ	State	Input	Output	Train A	Train B
0	SAFE	(0,0,0,0)	(1,1)	safe	safe
1	SAFE	(0,0,1,0)	(1,1)	safe	S3
2	SAFE	(0,0,0,0)	(1,1)	safe	danger
3	SAFE	(1,0,0,0)	(1,1)	S1	danger
4	AinBsafe	(1,0,0,0)	(1,1)	S1	danger
5	AinBsafe	(1,0,0,1)	(1,1)	S1	S4
6	AinBsafe	(1,0,0,0)	(1,1)	S1	safe
7	AinBsafe	(0,0,0,0)	(1,1)	danger	safe
8	AinBsafe	(0,1,0,0)	(1,1)	S2	safe
9	SAFE	(0,0,0,0)	(1,1)	safe	safe
10	SAFE				

Test Trajectory 3

TZ	State	Input	Output	Train A	Train B
0	SAFE	(0,0,0,0)	(1,1)	safe	safe
1	SAFE	(1,0,0,0)	(1,1)	S1	safe
2	AinBsafe	(0,0,0,0)	(1,1)	danger	safe
3	AinBsafe	(0,0,1,0)	(1,1)	danger	S3
4	AinBoff	(0,0,1,0)	(1,0)	danger	S3
5	AinBoff	(0,1,1,0)	(1,0)	S2	S3
6	BinAsafe	(0,0,1,0)	(1,1)	safe	S3
7	BinAsafe	(0,0,0,0)	(1,1)	safe	danger
8	BinAsafe	(0,0,0,1)	(1,1)	safe	S4
9	SAFE	(0,0,0,0)	(1,1)	safe	safe
10	SAFE	(0,0,0,0)	(1,1)	safe	safe
11	SAFE	(0,0,1,0)	(1,1)	safe	S3
12	SAFE	(0,0,0,0)	(1,1)	safe	danger
13	SAFE	(1,0,0,0)	(1,1)	S1	danger
14	AinBsafe	(1,0,0,0)	(1,1)	S1	danger
15	AinBsafe	(1,0,0,1)	(1,1)	S1	S4
16	AinBsafe	(1,0,0,0)	(1,1)	S1	safe
17	AinBsafe	(0,0,0,0)	(1,1)	danger	safe
18	AinBsafe	(0,1,0,0)	(1,1)	S2	safe
19	SAFE	(0,0,0,0)	(1,1)	safe	safe
20	SAFE				

Test Trajectory 4

TZ	State	Input	Output	Train A	Train B
0	SAFE	(0,0,0,0)	(1,1)	safe	safe
1	SAFE	(0,0,1,0)	(1,1)	safe	S3
2	SAFE	(0,0,0,0)	(1,1)	safe	danger
3	SAFE	(1,0,0,0)	(1,1)	S1	danger
4	AinBsafe	(1,0,0,0)	(1,1)	S1	danger
5	AinBsafe	(1,0,0,1)	(1,1)	S1	S4
6	AinBsafe	(1,0,0,0)	(1,1)	S1	safe
7	AinBsafe	(0,0,0,0)	(1,1)	danger	safe
8	AinBsafe	(0,1,0,0)	(1,1)	S2	safe
9	SAFE	(0,0,0,0)	(1,1)	safe	safe
10	SAFE	(0,0,0,0)	(1,1)	safe	safe
11	SAFE	(1,0,0,0)	(1,1)	S1	safe
12	AinBsafe	(0,0,0,0)	(1,1)	danger	safe
13	AinBsafe	(0,0,1,0)	(1,1)	danger	S3
14	AinBoff	(0,0,1,0)	(1,0)	danger	S3
15	AinBoff	(0,1,1,0)	(1,0)	S2	S3
16	BinAsafe	(0,0,1,0)	(1,1)	safe	S3

Test Trajectory 4 *continued*

TZ	State	Input	Output	Train A	Train B
17	BinAsafe	(0,0,0,0)	(1,1)	safe	danger
18	BinAsafe	(0,0,0,1)	(1,1)	safe	S4
19	SAFE	(0,0,0,0)	(1,1)	safe	safe
20	SAFE				

Test Trajectory 5

TZ	State	Input	Output	Train A	Train B
0	SAFE	(0,0,0,0)	(1,1)	safe	safe
1	SAFE	(1,0,1,0)	(1,1)	S1	S3
2	AinBoff	(0,0,1,0)	(1,0)	danger	S3
3	AinBoff	(0,0,1,0)	(1,0)	danger	S3
4	AinBoff	(0,1,1,0)	(1,0)	S2	S3
5	BinAsafe	(0,0,1,0)	(1,1)	safe	S3
6	BinAsafe	(0,0,0,0)	(1,1)	safe	danger
7	BinAsafe	(0,0,0,1)	(1,1)	safe	S4
8	SAFE	(0,0,0,0)	(1,1)	safe	safe
9	SAFE				

End of test.

6.6.2   System functional analysis of Concept 2

6.6.2.1   Top level system functional analysis of Concept 2

Concept 2 specifies creating an assembly language program to implement the requirements. Based on Document 3 requirements and the states defined in Document 5 for Concept 2, the major functions are identical to those for Concept 1. Refer to Figure 6.6 for the top level state diagram.

6.6.2.2   Functional decomposition

6.6.2.2.1   *Function 1.*   The first function is Monitor13. Based on Document 3, the following conditions must be set:

1. Location outside of the danger zone. This means the initial input is IRP1 = (0,0,0,0).

2. Power ON to both trains. This means that the initial output is ORP1 = (1,1).

Therefore, the output from this function is power ON for both trains. This function is left if either Switch 1 is triggered (I1P1 = 1) or Switch 3 is triggered

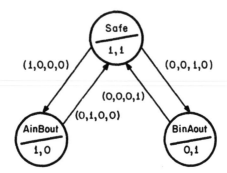

Figure 6.6 Top level state diagram for Alternative-2.

(I3P1 = 1). The assembly language program senses the inputs sequentially rather than in parallel, as is done in the protoboard. Because of this, it is helpful to break this function down into two subfunctions. The first function tests for S1 and the second tests for S3. If one of the tests fails the other function is entered. This testing continues, with the functions oscillating back and forth, until S1 or S3 is detected. This is a better model than that used for Concept 1 because the software tests for S1 and ignores all other switches. This prevents the system from entering an unknown state. The test of S1 will be called Monitor1 and the safe test of S3 will be called Monitor3. Power will remain ON to both trains while in these functions. These functions are patterned after states SAFE1 and SAFE2. See Figure 6.7.

   6.6.2.2.2 Function 2.   In the second function, Monitor23, Train A is in the danger zone and Train B is in the safe zone. This function is entered when Train A triggers S1 (I1P1 = 1) while in the Monitor13 function.

   To prevent collisions, the power to Train B is OFF in this function. To increase the I/O Performance Requirement for the number of laps Train B completes, it is desirable to decompose this function into two subfunctions that allow Train B to have power until it attempts to enter the danger zone. The function Monitor23 is decomposed into Monitor23On and Monitor23off. In addition, the model works better if Monitor23On is further decomposed into two functions, each of which tests a different switch. The function Monitor3On tests Switch 3 and Monitor2On tests Switch 2. Monitor3On enters the Monitor2On function if it does not see Switch 3 triggered. Monitor2On enters the Monitor14 function if it detects S2 triggered, otherwise it returns to the Monitor3On function. These functions are patterned after states AinBsafe1 and AinBsafe2. Refer to Figure 6.7 for the new state diagram. Monitor3On is left and Monitor3Off is entered if Train B triggers S3 [input (0,0,1,0) is detected]. Monitor 3Off goes to Monitor14 if S2 is triggered [input (0,1,0,0) is detected], otherwise it remains in that function.

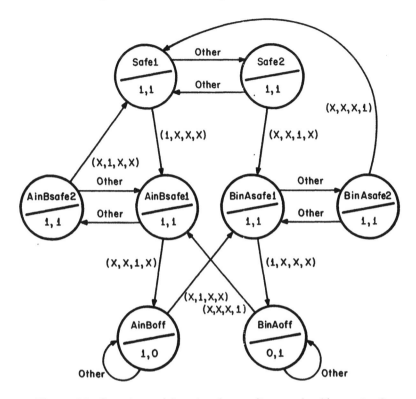

Figure 6.7   Complete subfunctional state diagram for Alternative-2.

6.6.2.2.3   Function 3.   In the third function, Monitor14, Train B is in the danger zone and Train A is in the safe zone. This function is entered when Train B triggers S3 (I3P1 = 1) while in the Monitor13 function.

To prevent collisions, power to Train A is OFF. To increase the Perform- ance Requirement for the number of laps Train A completes, it is desirable to decompose this function into two subfunctions that allow Train A to have power until it attempts to enter the danger zone. The Monitor24 function is decomposed into Monitor14On and Monitor14Off. In addition, the model works better if Monitor14On is further decomposed into two subfunctions, each of which tests a different switch. The function Monitor1On tests Switch 1 and Monitor4On tests Switch 4. Monitor1On enters the Monitor4On function if it does not see Switch 1 triggered. Monitor4On enters the Monitor13 function if it detects S4 triggered, otherwise it returns to the Monitor1On function. These are patterned after states BinAsafe1 and BinAsafe2. Refer to Figure 6.7 for the new state diagram. Monitor1On is left and Monitor1Off is entered if Train B triggers S1 [input (1,0,0,0) is detected]. Monitor1Off goes to Monitor13 if S4 is triggered [input (0,0,0,1) is detected], otherwise it remains in that function.

6.6.2.3   Complete functional model

6.6.2.3.1   Terminology

```
Z2' = (SZ2', IZ2', OZ2', NZ2', RZ2')
```

where

> $Z2'$ = the model of a complete I/O functional design,
> $SZ2'$ = states of system $Z2'$,
> $IZ2'$ = inputs to system $Z2'$,
> $OZ2'$ = outputs of system $Z2'$,
> $NZ2'$ = next state function, and
> $RZ2'$ = readout function.

6.6.2.3.2   States

```
SZ2' = {SAFE1,SAFE2,AinBsafe1,AinBsafe2,AinBoff,
        BinAsafe1,BinAsafe2,BinAoff}
```

6.6.2.3.3   Inputs

```
IZ2' = IZ1'
```

6.6.2.3.4   Outputs

```
OZ2' = OZ1'
```

6.6.2.3.5   Next state function

```
NZ2' = {(SAFE1,(1,0,0,0),AinBsafe1),
        (SAFE1,other,SAFE2),
        (SAFE2,(0,0,1,0),BinAsafe1),
        (SAFE2,other,SAFE1),
        (AinBsafe1,(0,0,1,0),AinBoff),
        (AinBsafe1,other,AinBsafe2),
        (AinBsafe2,(0,1,0,0),SAFE),
        (AinBsafe2,other,AinBsafe1),
        (AinBoff,(0,1,0,0),BinAsafe1),
        (AinBoff,other,AinBoff),
        (BinAsafe1,(1,0,0,0),BinAoff),
        (BinAsafe1,other,BinAsafe2),
        (BinAsafe2,(0,0,0,1),SAFE),
        (BinAsafe2,other,BinAsafe1),
        (BinAoff,(0,0,0,1),AinBsafe1),
        (BinAoff,other,BinAoff)}
```

where an input of "other" means any other valid input not specifically listed in the $NZ2'$ function.

6.6.2.3.6 *Readout function*

```
RZ2' = {(SAFE1,(1,1)),(SAFE2,(1,1)),
        (AinBsafe1,(1,1)),(AinBsafe2,(1,1)),
        (AinBoff,(1,0)),(BinAsafe1,(1,1)),
        (BinAsafe2,(1,1),(BinAoff,(0,1))}
```

6.6.2.4 *Rationale for analysis of Concept 2*

The Concept 2 functional decomposition is based on the requirements of Document 3 and the Concept 2 portion of Document 5. A computer simulation was run to test the complete functional model. Exhibit 6.3 is the simulation program listing, and Exhibit 6.4 is the result. All tests were passed, and the model satisfied all functional requirements.

EXHIBIT 6.3

Simulation Program Listing for Concept 2

```
/*PSEUDO CODE FOR SIMULATION OF THE SOFTWARE
              FUNCTIONAL DESIGN*/

/*----------------------------------------------*/


/*Beginning of loop for test trajectory*/
for each test trajectory do
   put power on to each train per test
       trajectory
   position each train per test trajectory

/*Beginning of loop for input data*/
   get next input of test trajectory until
       done
     if the current state is SAFE2
     /*SAFE2*/
         if the input is (x,x,1,x) then
         /* S3 */
             the next state is BinAsafe1
             B is in danger
         else
         /*other*/
             the next state is SAFE1
             B is safe
             A is safe
     else if the current state is SAFE1
```

```
/*SAFE1*/
   if the input is (x,x,1,x) then
   /* S1 */
      the next state is AinBsafe1
      A is in danger
   else
   /*other*/
      the next state is SAFE2
      B is safe
      A is safe
else if the current state is AinBsafe1
/*AinBsafe1*/
   if the input is (x,x,1,x) then
   /* S3 */
      the next state is AinBoff
         B has power off
         A is in danger
   else
   /*other*/
      the next state is AinBsafe2
      A is in danger
else if the current state is AinBsafe2
/*AinBsafe2*/
   if the input is (x,1,x,x) then
   /* S2 */
      the next state is Safe1
         A is safe
   else
   /*other*/
      the next state is AinBsafe1
      A is in danger
else if the current state is BinAsafe1
/*BinAsafe1*/
   if the input is (1,x,x,x) then
   /* S1 */
      the next state is BinAoff
         A has power off
         B is in danger
   else
   /*other*/
      the next state is BinAsafe2
      B is in danger
else if the current state is BinAsafe2
/*BinAsafe2*/
```

```
                 if the input is (x,x,x,1) then
                 /* S4 */
                    the next state is Safe1
                       B is in safe
                 else
                 /*other*/
                    the next state is BinAsafe1
                    B is in danger
           else if the current state is AinBoff
           /*AinBoff*/
                 if the input is (x,1,x,x) then
                 /* S2 */
                    the next state is BinAsafe1
                       B has power on
                       B is in danger
                       A is safe
                 else
                 /*other*/
                    the next state is AinBoff
                    B has power off
                    A is in danger
           else if the current state is BinAoff
           /*BinAoff*/
                 if the input is (x,x,x,1) then
                 /* S4 */
                    the next state is AinBsafe1
                       A has power on
                       A is in danger
                       B is safe
                 else
                 /*other*/
                    the next state is BinAoff
                    A has power off
                    B is in danger
           end of if loop for state
        if A is in danger and B is in danger then
/*OUTPUT ROUTINE*/
        print COLLISION
     else if A is in danger then
        print A IN DANGER
        print B IS SAFE
     else if B is in danger then
        print B IN DANGER
        print A IS SAFE
```

```
    else
        print A IS SAFE
        print B IS SAFE
        end of output routine if loop
    end loop for test trajectory input
end loop for test trajectory
```

6.6.3   System functional analysis of Concept 3

6.6.3.1   Top level system functional analysis of Concept 3
Concept 3 specifies creating a program using Pascal language software to implement the requirements. Based on Document 3 requirements and the states defined in Document 5 for Concept 3, the major functions are the same as those in Concepts 1 and 2.

6.6.3.2   Functional decomposition

6.6.3.2.1   *Function 1.*   This system is functionally the same as Concept 2 and decomposes into the same functions.

6.6.3.3   Complete functional model

6.6.3.3.1   *Terminology*

$$Z3' = (SZ3', IZ3', OZ3', NZ3', RZ3')$$

where

$Z3'$ = the model of a complete I/O functional design,
$SZ3'$ = states of system $Z3'$,
$IZ3'$ = inputs to system $Z3'$,
$OZ3'$ = outputs of system $Z3'$,
$NZ3'$ = next state function, and
$RZ3'$ = readout function.

6.6.3.4   Rationale for analysis of Concept 3
Concept 3 is functionally identical to Concept 2. They both specify software programs to be downloaded to the Motorola controller. Both the programs break down into the same functional areas. The simulation for Concept 3 will produce the same result as Concept 2.

EXHIBIT 6.4

Simulation Results for Concept 2

These tables represent the output of the simulation code in Exhibit 6.3. The input to the system was the test trajectories specified in the Test Requirement.

Begin test of software conceptual design simulation.

Test Trajectory 1

TZ	State	Input	Output	Train A	Train B
0	SAFE1	(0,0,0,0)	(1,1)	safe	safe
1	SAFE2	(1,0,0,0)	(1,1)	S1	safe
2	AinBsafe2	(0,0,0,0)	(1,1)	danger	safe
3	AinBsafe1	(0,0,1,0)	(1,1)	danger	S3
4	AinBoff	(0,0,1,0)	(1,0)	danger	S3
5	AinBoff	(0,1,1,0)	(1,0)	S2	S3
6	BinAsafe1	(0,0,1,0)	(1,1)	safe	S3
7	BinAsafe2	(0,0,0,0)	(1,1)	safe	danger
8	BinAsafe1	(0,0,0,1)	(1,1)	safe	S4
9	SAFE2	(0,0,0,0)	(1,1)	safe	safe
10	SAFE1				

Test Trajectory 2

TZ	State	Input	Output	Train A	Train B
0	SAFE1	(0,0,0,0)	(1,1)	safe	safe
1	SAFE2	(0,0,0,0)	(1,1)	safe	safe
2	BinAsafe1	(0,0,1,0)	(1,1)	safe	S3
3	BinAsafe2	(0,0,0,0)	(1,1)	safe	danger
4	BinAsafe1	(1,0,0,0)	(1,1)	S1	danger
5	BinAoff	(1,0,0,0)	(0,1)	S1	danger
6	AinBsafe1	(1,0,0,1)	(1,1)	S1	S4
7	AinBsafe2	(1,0,0,0)	(1,1)	S1	safe
8	AinBsafe1	(0,0,0,0)	(1,1)	danger	safe
9	AinBsafe2	(0,1,0,0)	(1,1)	S2	safe
10	SAFE1	(0,0,0,0)	(1,1)	safe	safe
11	SAFE2	(0,0,0,0)	(1,1)	safe	safe
12	SAFE1				

Test Trajectory 3

TZ	State	Input	Output	Train A	Train B
0	SAFE1	(1,0,0,0)	(1,1)	safe	safe'
1	SAFE2	(0,0,0,0)	(1,1)	safe	safe
2	SAFE1	(1,0,0,0)	(1,1)	S1	safe
3	AinBsafe1	(0,0,0,0)	(1,1)	danger	safe
4	AinBoff	(0,0,1,0)	(1,0)	danger	S3
5	AinBoff	(0,0,1,0)	(1,0)	danger	S3
6	BinAsafe1	(0,1,1,0)	(1,1)	S2	S3
7	BinAsafe2	(0,0,1,0)	(1,1)	safe	S3
8	BinAsafe1	(0,0,0,0)	(1,1)	safe	danger
9	BinAsafe2	(0,0,0,1)	(1,1)	safe	S4
10	SAFE1	(0,0,0,0)	(1,1)	safe	safe
11	SAFE2	(0,0,0,0)	(1,1)	safe	safe
12	BinAsafe1	(0,0,1,0)	(1,1)	safe	S3
13	BinAsafe2	(0,0,0,0)	(1,1)	safe	danger
14	BinAsafe1	(1,0,0,0)	(1,1)	S1	danger
15	BinAoff	(1,0,0,0)	(0,1)	S1	danger
16	AinBsafe1	(1,0,0,1)	(1,1)	S1	S4
17	AinBsafe2	(1,0,0,0)	(1,1)	S1	safe
18	AinBsafe1	(0,0,0,0)	(1,1)	danger	safe
19	AinBsafe2	(0,1,0,0)	(1,1)	S2	safe
20	SAFE1	(0,0,0,0)	(1,1)	safe	safe
21	SAFE2	(0,0,0,0)	(1,1)	safe	safe
22	SAFE1				

Test Trajectory 4

TZ	State	Input	Output	Train A	Train B
0	SAFE1	(0,0,0,0)	(1,1)	safe	safe
1	SAFE2	(0,0,0,0)	(1,1)	safe	safe
2	BinAsafe1	(0,0,1,0)	(1,1)	safe	S3
3	BinAsafe2	(0,0,0,0)	(1,1)	safe	danger
4	BinAsafe1	(1,0,0,0)	(1,1)	S1	danger
5	BinAoff	(1,0,0,0)	(0,1)	S1	danger
6	AinBsafe1	(1,0,0,1)	(1,1)	S1	S4
7	AinBsafe2	(1,0,0,0)	(1,1)	S1	safe
8	AinBsafe1	(0,0,0,0)	(1,1)	danger	safe
9	AinBsafe2	(0,1,0,0)	(1,1)	S2	safe
10	SAFE1	(0,0,0,0)	(1,1)	safe	safe
11	SAFE2	(0,0,0,0)	(1,1)	safe	safe
12	SAFE1	(1,0,0,0)	(1,1)	S1	safe
13	AinBsafe1	(0,0,0,0)	(1,1)	danger	safe
14	AinBoff	(0,0,1,0)	(1,0)	danger	S3
15	AinBoff	(0,0,1,0)	(1,0)	danger	S3

Test Trajectory 4 *continued*

TZ	State	Input	Output	Train A	Train B
16	BinAsafe1	(0,1,1,0)	(1,1)	S2	S3
17	BinAsafe2	(0,0,1,0)	(1,1)	safe	S3
18	BinAsafe1	(0,0,0,0)	(1,1)	safe	danger
19	BinAsafe2	(0,0,0,1)	(1,1)	safe	S4
20	SAFE1	(0,0,0,0)	(1,1)	safe	safe
21	SAFE2	(0,0,0,0)	(1,1)	safe	safe
22	SAFE1				

Test Trajectory 5

TZ	State	Input	Output	Train A	Train B
0	SAFE1	(0,0,0,0)	(1,1)	safe	safe
1	SAFE2	(0,0,0,0)	(1,1)	safe	safe
2	SAFE1	(1,0,1,0)	(1,1)	S1	S3
3	AinBoff	(0,0,1,0)	(1,0)	danger	S3
4	AinBoff	(0,0,1,0)	(1,0)	danger	S3
5	BinAsafe1	(0,1,1,0)	(1,1)	S2	S3
6	BinAsafe2	(0,0,1,0)	(1,1)	safe	S3
7	BinAsafe1	(0,0,0,0)	(1,1)	safe	danger
8	BinAsafe2	(0,0,0,1)	(1,1)	safe	S4
9	SAFE1	(0,0,0,0)	(1,1)	safe	safe
10	SAFE2	(0,0,0,0)	(1,1)	safe	safe
11	SAFE1				

End of test.

6.7   Document 7:   System Physical Synthesis

The System Physical Synthesis Document develops and explains the relationships between the models of the previous documents and the physical components that will comprise the final system. It is created in conjunction with Document 6.

6.7.1   Physical synthesis of Concept 1

6.7.1.1   Top level system design of Concept 1
Concept 1 is the hardware implementation of the train controller. TTL components are used to connect the train location detectors and the power controllers to the protoboard. The system is broken into the following physical units:

1. detector,
2. analyzer, and
3. output controller.

Each of these units is described in further detail in the following sections.

6.7.1.2   Subunit physical synthesis

6.7.1.2.1   Subunit 1.   Unit 1 is the detector. System inputs are scanned by this unit. It is implemented by providing four pins at the end of the protoboard in the following order: S1, S2, S3, and S4. The pins will interface to AND gates that connect to the analyzer. See Exhibit 6.5 for the state table, Figure 6.8 for the schematic, and Figure 6.9 for the track hardware.

6.7.1.2.2   Subunit 2.   Unit 2 is the analyzer. This unit takes the inputs from the detector and places them into three flip-flops. The state of all three flip-flops concurrently corresponds to the functional states of Section 6.6.1.2.2 to create the outputs. See Section 6.7.1.3 for the homomorphisms. This unit sends data to the output controller. See Figure 6.8 for the schematic.

6.7.1.2.3   Subunit 3.   Unit 3 is the output controller; it is used to set the power for the trains. It is implemented by connecting the output from the analyzer to two pins at the end of the protoboard. There are a total of eight pins on the protoboard. The first four are the inputs (see Section 6.7.1.2.1), the next two are earth ground (GND) and the last two are power to Train A (PA) and Train B (PB).

6.7.1.3   Homomorphisms
This section describes the relationships between the physical protoboard and the functional design of Concept 1 given in Document 6. HS is the state

EXHIBIT 6.5

State Table for Concept 1

Present State ABC(t)	Inputs S1 S2 S3 S4	Next State ABC(t + 1)	JA	KA	JB	KB	JC	KC
000	1 0 0 0	100	1	X	0	X	0	X
000	0 0 1 0	010	0	X	1	X	0	X
100	0 1 0 0	000	X	1	0	X	0	X
100	0 0 1 0	110	X	0	1	X	0	X
010	0 0 0 1	000	0	X	X	1	0	X
010	1 0 0 0	111	1	X	X	0	1	X
110	0 1 0 0	010	X	1	X	0	0	X
111	0 0 0 1	100	X	0	X	1	X	1

where X means "don't care."

$$JA = A'B'C'S1 + A'BC'S1 = A'C'S1(B + B') = A'C'S1$$

$$KA = AB'C'S2 + ABC'S2 = AC'S2(B + B') = AC'S2$$

$$JB = A'B'C'S3 + AB'C'S3 = B'C'S3(A + A') = B'C'S3$$

$$KB = A'BC'S4 + ABCS4$$

$$JC = A'BC'S1$$

$$KC = S4$$

This design seems to ignore the simultaneous occurrence of S1 and S3, and no mention is made of possible capture in unused states. Detailed analysis has shown that this circuit does overcome these problems.

These Boolean functions can be reduced to

$$JA = S1, \quad KA = S2, \quad JB = S3, \quad KB = S4,$$
$$JC = A'BC'S1, \quad KC = S4$$

but the circuit would no longer overcome these problems.

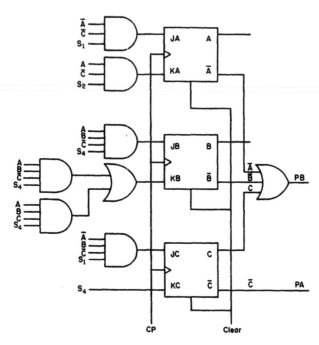

Figure 6.8 Schematic for Alternative-1.

homomorphism. It defines the state **SAFE** in Document 6, Section 6.6.1.3.2, as having the value 000 (binary) in the description of the protoboard (as specified in Document 7, Exhibit 6.5).

```
HS = {(SAFE,000),(AinBsafe,100),(AinBoff,010),
      (BinAsafe,110),(BinAoff,111)}
```

where the state 000 corresponds to the three flip-flops having outputs of 0 in the order A, B, C.

   HI is the input homomorphic relationship relating the input **S1** in Document 7 to **I1Z1** in Document 6, Section 6.6.1.3.3. Similarly, **HO** is the output homomorphism relating **PA** from Document 7 to **01Z1** from Document 5, Section 6.6.1.3.4.

```
HI = {(S1,I1Z1),(S2,I2Z1),(S3,I3Z1),(S4,I4Z1)}
HO = {(PA,01Z1),(PB,02Z1)}
```

6.7.1.4 *Rationale for synthesis of Concept 1*
The design shown in Figure 6.8 requires five integrated circuits (ICs). Two ICs are for the flip-flops and three are for the AND and OR gates. This design approach was derived using the state diagram created from Document 6, Functional Analysis. Many other TTL designs are certainly possible.

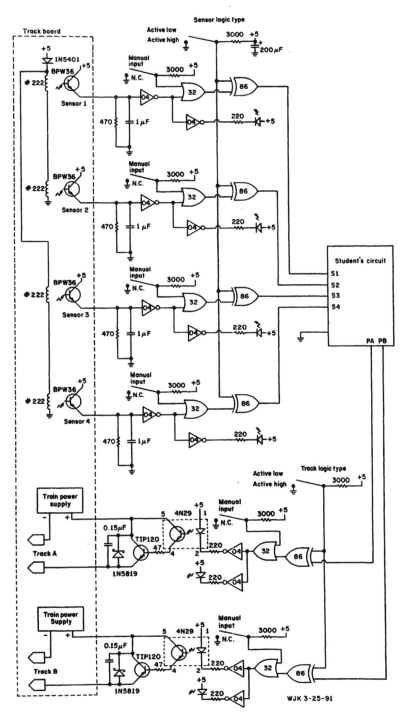

Figure 6.9 Schematic for the electronics interface to the track hardware.

Designing for manufacturability is essential—the system design must consider the manufacturing processes. In line with this statement, we want to minimize the number of integrated circuits, discrete components (resistors, capacitors, LEDs, etc.), and interconnecting wires in SIERRA. But how should each of these be evaluated? To help answer this question the manufacturing department should be consulted early in the design process. Knowing the approximate size of the project, they can suggest, for example, wire wrap, single-sided printed circuit boards, multilayer circuit boards, etc. Based on this decision, they can determine the relative expense of integrated circuits, discrete components, and interconnecting wires.

6.7.2   Physical synthesis of Concept 2

6.7.2.1   Top level system design of Concept 2

Concept 2 uses an assembly language program to implement the train controller. The physical unit is the Motorola controller available in the lab. It will be broken down into two units:

1. input/output and
2. analyzer

Each is explained below.

6.7.2.2   Subunit physical synthesis

6.7.2.2.1   *Subunit 1.*   The input/output unit is a parallel port on the Motorola controller. The TA will connect the equipment to the controller. This unit presents input/output data in memory location $10010 (where $ indicates a binary number) of the controller's memory. Bits 1,2,3, and 4 of the port correspond to S1, S2, S3, and S4. Bits 5 and 6 correspond to PA and PB.

6.7.2.2.2   *Subunit 2.*   Unit 2 is the analyzer, which consists of the Motorola controller available in the SIE-370 lab and its downloaded software. It is required equipment as per the Operational Need Document. Software written in the Motorola assembly language must be downloaded to this unit. The software will be written on a Unix-based PC or DOS-based IBM PC in the lab in assembly language, then it will be downloaded to the Motorola controller. The analyzer software can be further decomposed to follow the functions described for Concept 2 in the Functional Analysis Document. This has been done in the flow diagram (see Figure 6.10). Exhibit 6.6 is the software listing for this flow diagram.

6.7.2.3 *Homomorphisms*

This section describes the relationships between the physical software and the functional system for Concept 2 as described in Document 6.

```
HS = {(a line 3,SAFE1),(a line 5,SAFE2),
      (c line 1,AinBsafe1),(c line 3,AinBsafe2),
      (b line 1,BinAsafe1),(b line 3,BinAsafe2),
      (d,AinBoff),(e,BinAoff)}
```

In this case a, b, c, d, and e refer to lines in the software.

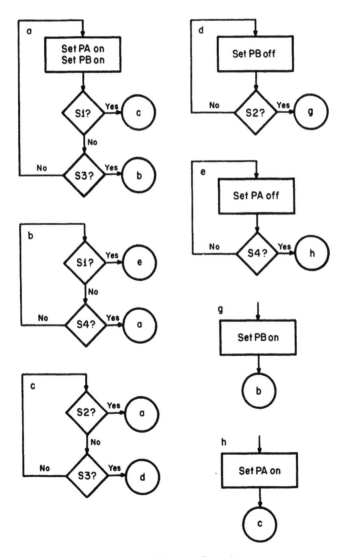

Figure 6.10 Software flow diagram.

EXHIBIT 6.6

Assembly Language Software Listing

```
      org 0x1000               * initialization work
      mov.b &0,1(%a0)          * initialization work
      mov.b &0,3(%a0)          * initialization work
      mov.b &0x80,15(%a0)      * initialization work
      mov.b &0x60,7(%a0)       * initialization work
      lea 0x10013,%a0          * initialization work
a:
      bset &5,(%a0)    * turn on power of train A
      bset &6,(%a0)    * turn on power of train B
      btst &1,(%a0)    * check for S1 (bit 1)
      bne c            * if S1 then goto c
      btst &3,(%a0)    * if no S1 then check for S3
      bne b            * if S3 then goto b
      bra a            * if no S3 then goto a
c:
      btst &2,(%a0)    * check for S2 (bit 2)
      bne a            * if S2 then goto a
      btst &3,(%a0)    * if no S2 then check for S3
      bne d            * if S3 then goto d
      bra c            * if no S3 then goto c
b:
      btst &1,(%a0)    * check for S1 (bit 1)
      bne e            * if S1 then goto e
      btst &4,(%a0)    * if no S1 then check S4
      bne a            * if S4 then goto a
      bra b            * if no S4 then goto b
d:
      bclr &6,(%a0)    * turn power of train B OFF
      btst &2,(%a0)    * check for S2 (bit 2)
      bne g            * if S2 then goto g
      bra d            * if no S2 then goto d
e:
      bclr &5,(%a0)    * turn power to train A OFF
      btst &4,(%a0)    * check for S4 (bit 4)
      bne h            * if S4 then goto h
      bra e            * if no S4 then goto e
g:
      bset &6,(%a0)    * turn power of train B ON
      bra b            * goto b
```

```
h:
    bset &5,(%a0)      * turn power to train A ON
    bra c              * goto c
    org 0x1000             * initialization work
    mov.b &0,1(%a0)        * initialization work
    mov.b &0,3(%a0)        * initialization work
    mov.b &0x80,15(%a0)    * initialization work
    mov.b &0x60,7(%a0)     * initialization work
    lea 0x10013,%a0        * initialization work
```

```
HI = {($10010 BIT 1,I1Z1),($10010 BIT 2,I2Z1),
      ($10010 BIT 3,I3Z1),($10010 BIT 4,I4Z1)}
HO = {($10010 BIT 5,O1Z1),($10010 BIT 6,O2Z1)}
```

6.7.2.4 Rationale for synthesis of Concept 2

Concept 2 was derived from the Technology Requirements in Document 3, the Concept Exploration Document and the Functional Analysis Document. The decomposition into functions specified in the Functional Analysis Document for Concept 2 was easily accomplished using the software.

6.7.3 Physical synthesis of Concept 3

6.7.3.1 Top level system design of Concept 3

When the physical synthesis for Concept 3 was begun, it was discovered that no high-level language cross compiler for the Unix-based PC or the DOS-based IBM PC was available for the Motorola controller. As a result of this, Concept 3 was not completed.

Problems

1. **Buildability-1.** This problem concerns the hardware controller for SIERRA designed in Document 5. The following describes TKY1, the technology you are allowed to use:

TTL integrated circuits
2.2 kΩ resistors
500 Ω resistors
simple switches
22 gauge wire
protoboard
light emitting diodes (LEDs)

Is this system buildable using technology TKY1?

2. **Buildability-2.** This problem also concerns the hardware controller for SIERRA designed in Document 5. The following describes TKY2, the technology you are allowed to use:

TTL integrated circuits   ˈ
2.2 kΩ resistors
500 Ω resistors
simple switches
22 gauge wire
protoboard
light emitting diodes (LEDs)
a three wire power cord with a three pronged plug
any electrical receptacle in the room

Is this system buildable using technology TKY2?

3. † **Buildability-3.** This problem also concerns the hardware controller for SIERRA designed in Document 5. The following describes TKY3, the technology you are allowed to use:

the following TTL integrated circuits: 7404, 7408, 7421, 7432, 7473
2.2 kΩ resistors
500 Ω resistors
simple switches
22 gauge wire
protoboard
light emitting diodes (LEDs)
555 timers
resistors and capacitors for the timer circuits
a 5 V battery system

Is this system buildable using technology TKY3?

4. **Buildability-4.** Let $Z41$ be defined as follows:

```
Z41 = (SZ41, IZ41, OZ41, NZ41, RZ41),
```

where

```
SZ41 = {1, 2} × {11, 12},
IZ41 = {3, 4},
OZ41 = {15, 16},
NZ41 = {((((1, 11), 3), (1, 11)),
        (((1, 11), 4), (2, 11)),
        (((1, 12), 3), (1, 12)),
        (((1, 12), 4), (2, 12)),
```

---

† This problem requires technical material not presented in this text.

```
              (((2, 11), 3), (2, 12)),
              (((2, 11), 4), (1, 12)),
              (((2, 12), 3), (2, 11)),
              (((2, 12), 4), (1, 11))},
   RZ41 = {(((1, 11), 15), ((1, 12), 16),
              ((2, 11), 15), ((2, 12), 16)}
```

Show that $Z41$ is buildable using TKYx4, where TKYx4 is the set of all systems that are isomorphic to $Z4$ as given below. (The definition of isomorphic images was given in the solutions to Problems 18 and 22 in Chapter 3.)

```
   Z4 = (SZ4, IZ4, OZ4, NZ4, RZ4),
```

where

```
   SZ4 = {A, B},
   IZ4 = {Yes, No},
   OZ4 = {On, Off},
   NZ4 = {((A, No), A), ((A, Yes), B),
          ((B, No), B), ((B, Yes), A)},
   RZ4 = {(A, Off), (B, On)}.
```

5. † **Buildability-5.** A new combination lock has recently been installed on the door of our laboratory. It has five buttons that can be pressed individually or in combination and a door knob. Presume that the correct combination is to first push buttons 4 and 2 at the same time, followed by button 3. Turning the door knob resets the process.

A dialogue is necessary between the systems engineer and the client to educate the systems engineer about the purpose of the system. Below we provide you, the systems engineer, with a system design that will satisfy the Input/Output Functional Requirements you are supposed to come up with.

Note: In the following description, DC means don't care.

```
   Z12 = (SZ12, IZ12, OZ12, NZ12, RZ12),
```

where

```
   SZ12 = {Start, Incorrect-sequence, 4-and-2,
           Correct-sequence},
   I1Z12 = {0, 1}, /*0 means Button 1 not pushed,
                      1 means Button 1 pushed*/
   I2Z12 = {0, 1}, /*0 means Button 2 not pushed,
                      1 means Button 2 pushed*/
   I3Z12 = {0, 1}, /*0 means Button 3 not pushed,
                      1 means Button 3 pushed*/
```

---

† This problem requires technical material not presented in this text.

```
I4Z12 = {0, 1}, /*0 means Button 4 not pushed,
                   1 means Button 4 pushed*/
I5Z12 = {0, 1}, /*0 means Button 5 not pushed,
                   1 means Button 5 pushed*/
I6Z12 = {0, 1}, /*0 means knob not turned,
                   1 means knob turned*/
0Z12 = {Locked, Unlocked},
NZ12 = {((Start, (0, 1, 0, 1, 0, 0)), 4-and-2),
   ((Start, (DC, DC, DC, DC, DC, 1)), Start),
   ((Start, (any other inputs)),Incorrect-sequence),
   ((Incorrect-sequence,
     (DC, DC, DC, DC, DC, 1)), Start),
   ((Incorrect-sequence, (any other inputs)),
     Incorrect-sequence),
   ((4-and-2, (0, 0, 1, 0, 0, 0)),Correct-sequence),
   ((4-and-2, (DC, DC, DC, DC, DC, 1)), Start),
   ((4-and-2, (any other inputs)),
     Incorrect-sequence),
   ((Correct-sequence,
     (DC, DC, DC, DC, DC, 1)), Start),
   ((Correct-sequence, (any other inputs)),
     Incorrect-sequence)},
RZ12 = {(Start, Locked),
        (Incorrect-sequence, Locked),
        (4-and-2, Locked),
        (Correct-sequence, Unlocked)}.
```

Specify the Input/Output and Functional Requirements for such a system. Show at least three difference input trajectories.

Define technology TKYx5 as the set of all systems that have two states, two inputs, and two outputs, such as Z4 in Problem 4 above. Is this solution buildable or implementable in technology TKYx5? Why, or why not?

Definition: A system may not be buildable in a certain technology, yet it might still be implementable in that technology. A system Z1 is implementable in a certain technology if and only if there is another system Z2 that is buildable in that technology and that is able to simulate Z1.

---

6. **Scoring functions.** Sketch the scoring functions for the following three Performance Figures of Merit: Number of Collisions, Trips by Train A, and Availability. Also, sketch the scoring functions for the following two Utilization of Resources Figures of Merit: Cost, Quantity 1 and Cost, Quantity 1,000,000.

7. **Technology restrictions.** Discuss the qualitative differences in the design of two law enforcement systems, the technology of one specifically excluding firearms and the other permitting them.

chapter seven

Other system design tools

In this chapter, we discuss other tools and techniques used by system designers. The sections of this chapter are independent of one another and need not be examined in order.

7.1 Quality function deployment (QFD)

Over the past 40 years, the Japanese have developed many quality improvement techniques for manufacturing processes. One of these, quality function deployment (QFD), is becoming very popular in both Japan and the United States. QFD was developed in Japan in the late 1960s and is now used by half of Japan's major companies. Automobile manufacturing companies in the United States began using the tool in the early 1980s. Now many major corporations use it, including John Deere, Ford, Chrysler, General Motors, Hughes Aircraft, Boeing, McDonnell Douglas, Martin Marietta, Texas Instruments, and 3M.

The objective of QFD is to get the idea of quality introduced into the early phases of the design cycle and maintained throughout the entire life cycle of the product. In most implementations, QFD requires the use of many matrix-like charts (called Houses of Quality) to discover interrelationships among customer demands, engineering requirements, and manufacturing processes, as shown in Figure 7.1. For example, the first QFD chart compares the customer demands to quality characteristics. The second chart investigates the relationships between these quality characteristics and characteristics of the product. The third chart looks for relationships between product characteristics and characteristics of the manufacturing system. Finally, these manufacturing characteristics are related to the quality controls that will be monitored during manufacture. This process of studying interrelationships may continue through dozens of charts.

The Japanese concept is that everyone should participate in making the product better. (This sounds like a renaming of systems engineering.) Therefore, QFD data is presented in a user-friendly format—all system design tools should be usable by the Ph.D. chief scientist as well as the janitor lacking a high school diploma. Thus, QFD tools must be mathematically simple.

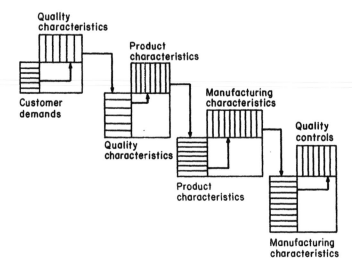

Figure 7.1 QFD waterfall chart.

The QFD approach begins with interviews of customers to elicit those characteristics most important to them, such as low cost, durability, comfort, etc. These characteristics, called Customer Demands in the QFD literature, are not specifications, they are general characteristics of the desired product. The customer must assign a weight to each demand indicating its relative importance. The range of weights is usually 1 to 10, with 10 being the most important. Figure 7.2 shows the Customer Demands and their associated weights for the Pinewood Derby study of Chapter 5.

To prove that Customer Demands are being satisfied, engineers must determine how to measure the Demands. In the QFD literature, these measures are usually called Quality Characteristics; in earlier chapters of this book (and generally in systems theory) we called them figures of merit. In general, QFD charts have a list of items on the left, which we call the Whats, and a list of items along the top, which we call the Hows (see Figure 7.3). To help determine the Hows, we ask the question, "This is *what* the customer wants, but *how* will it be measured?"

The next step in a QFD analysis is the determination of the strengths of the relationships (or the degree of correlation) between the Whats and the Hows. This is done by filling in the matrix, as shown in Figure 7.3. Each element of the Whats is evaluated with regard to each element of the Hows and a classification according to strength of relationship is assigned. In the figure, one of four classifications is assigned to every intersection: 9 for a strong relationship (a circle with a dot inside), 3 for a medium relationship (an open circle), 1 for a weak relationship (a triangle), and 0 for no relationship (the cell is left blank). Each relationship can be either positive or negative, but more important is a determination of whether or not each Customer Demand

	Weight
Lots of races per scout	7
Very few ties	5
Happy scouts	10
No irate parents	10
No broken cars	7
Nobody touch scout's car	4
Fair races	
Every scout races every scout	5
Every scout races in every lane	5
All race about same # times	4
Don't waste time	6
Easily implementable	9
Affordable	5

Figure 7.2 Customer Demands with their weights.

WHATs vs HOWs

Strong relationship: ⊙ 9
Medium relationship: O 3
Weak relationship: △ 1

WHATs vs HOWs	Number of races per scout	Observed scout behavior	Observed parent behavior	Number of broken cars	Number of touches/car/race	Results of every race	Lane	Car	Place	Length of division races	Training time	Judging resolution	Cost	Money	Time
Lots of races per scout	⊙	O	△	△	O			O		⊙	△			△	O
Very few ties		O	⊙		△					⊙	△	O	⊙		△
Happy scouts	⊙	⊙	O	△	O					△	△	⊙			
No irate parents		O	⊙	△			△			△	△	⊙			
No broken cars	△	△	△	⊙	O					△	△				
Nobody touch scout's car	△	O		△	⊙					△	△				
Fair races															
Every scout races every scout	△	O						⊙		△	△		△		
Every scout races in every lane	△	O	O				⊙			△	△		△		
All race about same # times	⊙	△	△						⊙	△			△		
Don't waste time	⊙	△	O	O						⊙	△				⊙
Easily implementable	⊙	△	△				△	△	△	⊙	⊙	O	O	O	O
Affordable												O		⊙	O

Figure 7.3 Customer Demands versus Quality Characteristics.

WHATs vs HOWs

Strong relationship: ◎ 9
Medium relationship: O 3
Weak relationship: △ 1

WHATs vs HOWs	Number of races per scout	Observed scout behavior	Observed parent behavior	Number of broken cars	Number of touches/car/race	Results of every race	Lane	Car	Place	Length of division races	Training time	Judging resolution	Cost	Money	Time	Weight
Lots of races per scout	◎	O	△	△	O			O		◎	△			△	O	7
Very few ties		O	◎		△			◎	△	O	◎				△	5
Happy scouts	◎	◎	O	△	O					△	△	◎				10
No irate parents		O	◎	△			△			△	△	◎				10
No broken cars	△	△	△	◎	O					△	△					7
Nobody touch scout's car	△	O		△	◎					△	△					4
Fair races																
Every scout races every scout	△	O						◎		△	△			△		5
Every scout races in every lane	△	O	O				◎			△	△			△		5
All race about same # times	◎	△	△					◎		△				△		4
Don't waste time	◎	△	O	O						◎	△				◎	6
Easily implementable	◎	△	△				△	△	△	◎	◎	O		O	O	9
Affordable												O		◎	O	5
Score	345	224	225	112	113		64	111	54	248	150	267		93	122	
Units	Races	% happy	% irate	Cars broken	Touches		Difference	Names	1st, 2nd, 3rd	Hours	Minutes	msec		Dollars	Days	
Target value	6	95	5	1	2					1	5	0.1		150		

Figure 7.4 Calculated scores and target values.

can be measured by a Quality Characteristic. If any row of this matrix is blank, satisfaction of that Demand cannot be assured. That Customer Demand should either be eliminated (not a good idea) or another Quality Characteristic added.

Next we multiply the value of each cell by the weight of the Customer Demand and total the column for each Quality Characteristic. This is done in the row labeled Score in Figure 7.4. The total score for each column is an indication of the importance of that Characteristic in measuring the customer's satisfaction. Measures with low scores typically receive little consideration.

However, this does not necessarily mean that these measures will not be used in the product design; they may still be necessary for contractual or other reasons. To meet the goal of satisfying the customer, we must be sure to pay strict attention to the measures with the highest scores. Attention to the customer's wants is the ultimate purpose of the QFD chart. The chart and its particular results are not as important as the entire process of concentrating on the "voice of the customer" rather than on the "voice of the producer." In the Pinewood Derby (see Figure 7.4), the Number of Races per Scout (with a score of 345) and the Judging Resolution (with a score of 267) were the most important measures. This makes sense because the scouts enjoy racing their cars and the crowd gets upset when an apparent winner is declared a loser. Target Values and their Units are also included in QFD charts (see Figure 7.4) so that engineers can design to these optimal values. The Target Values are those values that gave a score of 1.0 for the scoring functions defined in Chapters 5 and 6. The technique demonstrated in Chapters 5 and 6 has a big advantage over QFD because of its use of scoring functions. So far no one has incorporated scoring functions into QFD charts.

In addition to the relationships between the Whats and the Hows, Figure 7.5 shows correlations between the Hows in the top triangle, which is called the "roof" of the House of Quality. There are five possible levels of relationship between the Quality Characteristics: 9 for strong positive (a circle with a dot inside), 3 for weak positive (an open circle), 0 for none (a blank square), –3 for weak negative (a × symbol), and –9 for strong negative (a # symbol). In some of our examples we will use the symbols, and in others we will use the numbers. You should use whatever method your customers are most comfortable with. Different symbols may even be used. As stated by Akao (1990), the foremost principle of QFD is "copy the spirit, not the form."

Relationships between the Hows help to identify correlations between the measures. For example, the Number of Races per Scout is strongly related to the Length of Division Races. As one measure increases, the other will also increase. Another example is Observed Scout Behavior and Number of Broken Cars. This is a strong negative correlation because as the number of broken cars increases, the scouts behavior becomes worse (i.e., the measure of scout behavior is low when the scouts are not happy). We will explain the left portion of Figure 7.5 in a later section of this chapter.

The Quality Characteristics from the top of Figure 7.4 are used as entries in our second QFD chart, which is shown in Figure 7.6 (note that the entries are on the left side of this chart). This chart shows the Quality Characteristics versus the Product's Parts (or Components or Subsystems, whatever designation is most appropriate for the task at hand). The score of each Quality Characteristic was determined in the first chart and is used as the weight in the second chart. The Quality Characteristics become the new Whats and the Parts become the new Hows. The question now is, "This is *what* I am going to measure, but *how* will I build the part to optimize this?" It is sometimes helpful

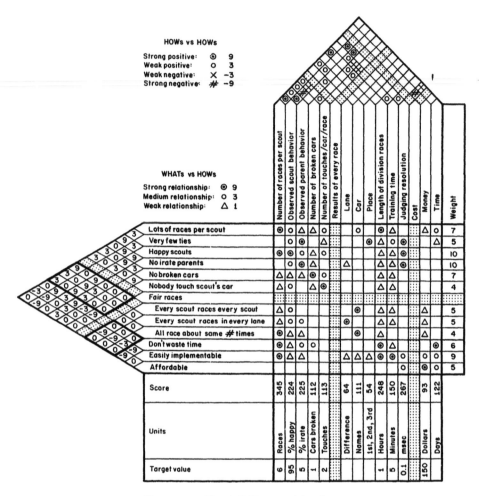

Figure 7.5   The full House of Quality.

to put several parts on the chart (or even the products of the competition) and compare their overall scores. This may help determine the trade-offs. We fill out this chart using the the same process used for the first chart. Fill out each cell based on how well the Part is related to the Quality Characteristics. Multiply the weights (the scores from the previous chart) by the numerical values for the relationships and sum the numbers in each column to give the scores at the bottom of this chart. The column scores now indicate how well each Part helps meet the Customer's Demands. For the Pinewood Derby, these scores indicate that Electronic Judging and Round Robin (Best Time) are the best mix of parts for our system design. This is the same result derived in the trade-off study of Document 5 in Chapter 5.

WHATs vs HOWs	Race format	Single elimination	Double elimination	Round robin point total	Round robin best	Round robin average time	Judging format	Human judging	Electronic judging	Weight	Units	Target value
Number of races per scout		△	o	◎	◎	◎		△		345	Races	6
Observed scout behavior			△	△	o	△		△	o	224	% happy	95
Observed parent behavior			△	△	△	△		△	◎	225	% irate	5
Number of broken cars		△	△							112	Cars broken	1
Number of touches/car/race		△	△							113	Touches	2
Results of every race												
Lane			△	◎	◎	◎				64	Difference	
Car		△	o	◎	◎	◎				111	Names	
Place		△	△					o	◎	54	1st, 2nd, 3rd	
Length of division races		o	o					△	o	248	Hours	1
Training time		o	o	△	△	△			△	150	Minutes	5
Judging resolution								o	◎	267	msec	0.1
Cost												
Money		o	o					◎		93	Dollars	150
Time		o	o					◎		122	Days	
Score		2574	3999	5279	5727	5279		3940	6480			

Strong relationship: ◎ 9
Medium relationship: o 3
Weak relationship: △ 1

Figure 7.6 The second QFD matrix, Quality Characteristics versus Parts.

The third QFD chart (not shown) evaluates the Parts with regard to the Manufacturing Processes. The goal here is to concentrate on the processes that are ultimately the most important to the customer. For example, a new car buyer views as critically important the process of painting the car's body, but not the process of painting the car's oil pan.

This technique of linking QFD charts together can continue until dozens of charts have been filled out (see Re Velle, 1988; King, 1989; and Akao, 1990), as suggested by the "waterfall" chart of Figure 7.1. King (1989) provides an example of linking dozens of QFD charts for one small heuristic example. Akao (1990), arguably the definitive work on QFD in the English language, gives many examples derived from real manufacturing systems. QFD can also be applied to the top-level function, then to subfunctions, and finally to their subfunctions. Using QFD to design real systems will involve the construction of many, many QFD charts. Computer programs, such as QFD/Capture (*QFD/Capture User's Manual*, 1990), are available to assist in managing the large databases generated by QFD analysis.

Creating QFD charts is a lot of work. A real productivity gain results from reusing parts of QFD charts that are already completed. If several versions of the same product are being built, parts of earlier QFD charts can be used in the design of later products in the line. QFD charts can also be reused for a product redesign.

The important thing to remember about QFD is that its goal is to satisfy the customer by manufacturing a product that incorporates those things the customer considers important. It is of little value to create many charts or to try to optimize a particular chart as an end in itself. The objective of QFD should be the discovery of what is important to the customer. Akao (1990) says, "The process of making QFD charts is more meaningful than the final results."

Study Figures 7.5 and 7.6 in detail. Look at each evaluation we made and decide if you would give it the same value. Change some of the values and see if the final recommendations are changed. You should now be prepared for the QFD homework problems.

7.1.1 Lessons learned making QFD charts

We will summarize a few of the lessons we have learned in creating QFD charts in the hope that this will help you fill out your charts. We will focus our attention on the original part, the "porch," of the House of Quality (the left, triangular part of Figure 7.5). In this chart, the values used for the Hows versus Hows relationships on the roof are the same as are used for the Whats versus Whats relationships on the porch. However, symbols are used on the roof and numbers are used on the porch.

Principles of psychology state that humans best understand those properties stated in a positive manner (avoiding negative statements) and those properties chosen so that "more is better" or so that "an optimum is desired" (i.e., so that scoring functions like those in Figures 4.2a and 4.2c can be used). For the Pinewood Derby, we used our customer's words to name our Whats, though this sometimes made our work more difficult. For example, in the House of Quality of Figure 7.5, the porch showed that the correlation between Don't Waste Time and Easily Implementable was negative. This means that as the system gets simpler there is less time *not* being wasted, which means that more time is wasted. It would have made more sense if we rephrased our Customer's Demands in a positive manner, with an entry such as Use Time Efficiently. For an example of confusion caused by ignoring the "more is better" maxim, look at the relationship between Very Few Ties and Easily Implementable. The correlation value is negative, which means that as the system gets simpler we can expect more ties.

Negative correlations can be confusing, but we do not want to simply eliminate all of them, because negative correlations point out conflicting customer demands that will make optimization difficult or perhaps make model validation impossible. For example, Every Scout Race in Every Lane is related negatively with Easily Implementable. We know that in this instance our customer is making conflicting demands. The trade-offs necessitated by these conflicting demands were one of the most important parts of our trade-off study in Chapter 5.

In assessing correlations, tertiary links should be avoided. For example, if Every Scout Races Every Scout then there must be Lots of Races Per Scout. And if there are Lots of Races Per Scout, there will surely be more Broken Cars. However, since the link between Every Scout Races Every Scout and No Broken Cars is only a tertiary link, we were careful not to indicate a correlation between these two in the porch of Figure 7.5.

Analyzing correlations in the porch of the House can help organize the Whats into appropriate subcategories. Whats that have similar correlations with the other Whats should be grouped together. For example, we grouped the three customer demands Every Scout Races Every Scout, Every Scout Races in Every Lane, and All Race About Same # Times into one What called Fair Races. We did this because of the similarity of the numbers in the three diagonal rows correlating these three demands with all the others. (These three rows are outlined in bold in the porch of Figure 7.5.) We do not want to eliminate some of these demands, because the three demands are independent of one another, as can be seen by the triangle of zeroes between these three diagonal rows.

Let us consider a different example to further illustrate the need to use the porch to help group similar entries. Suppose a young couple wants to buy a new car. The man says that his most important demand is Horse Power and the woman says that her most important demand is Gas Mileage. Although these are conflicting demands with a negative correlation, there should be no problem. Their decision of what car to buy will probably be based on a trade-off between these two criteria. Now assume that a different couple wants to buy a car. Like the first woman, this woman says that her most important demand is Gas Mileage, but the man says that his most important demands are Lots of Horse Power, Lots of Torque, Low Time to Accelerate 0 to 60 mph, Low Time to Accelerate 0 to 100 mph, Low Time for the Standing Quarter Mile, Large Engine Size (in liters), and Many Cylinders. The man agrees that the woman's demands are more important than his, so they decide to weight Gas Mileage the heaviest, giving it the maximum importance value of 10. They give the man's demands importance values of 3 and 4. What kind of car do you think they will buy? In summary, dependent entries should be combined, but similar, though independent, entries should be made into subcategories and grouped together.

Every QFD chart can have two correlation matrices attached; we have called these the porch and the roof. In all cases, they alert the system designer to interactions that have consequences the nature of which depend on the particular QFD chart. Consider a correlation matrix in which the system components are listed. In a system that is assembled from components made by different people, divisions, or companies, it is important to know which components effect other components. Thus, a division that makes a change in the component being built can notify the other divisions that will be affected by the change.

A valuable principle in studying correlations is "Do not let preconceived notions about the solution affect the treatment of the customer's demands." For example, in all our alternative designs, the simpler solutions (such as Human Judges) led to more ties. Therefore, when we filled in the QFD charts, we said there was a correlation. However, this correlation only exists because of our preconceived notions, not because of the customer's stated demands. There might indeed be a very simple solution that precludes ties that we just did not think of.

7.1.2   Other QFD charts

Figure 7.1 shows a temporal ordering of QFD matrices. It would be nice if we really could design a product in such a straightforward manner. Unfortunately, more often than not everything has to be done simultaneously. We now briefly discuss some of the other types of QFD charts that are used. Most of them do not follow a temporal order. The Whats and Hows used in these charts may include: Customer Demands, Quality Characteristics, Alternatives, Functions, Parts, Components, Mechanisms, Product Failure Modes, Part Failure Modes, and New Concepts. With just these ten entities, over one million matrices can be formed. However, not all of these matrices are useful—less than 100 are in common use. King (1989) explains 30 of these QFD charts. Below we state the basic purpose of a few of them.

(1) *Customer Demands versus Quality Characteristics:* This is the House of Quality of Figure 7.4. Purposes: to help the system designers learn customer priorities, to point out which Customer Demands are the most important, to ensure that no Customer Demand is ignored, to identify key items to measure and control, and to help the designers develop an initial plan of how Customer Demands will be satisfied. This is the most widely used QFD chart.

(2) *Customer Demands versus Customer Demands:* This is the porch of the House in Figure 7.5. It can be constructed as a separate chart. Purpose: to alert the system designers to interactions and to assist in eliminating dependent Demands and in grouping similar but independent Demands into subcategories. We present this chart here for the first time; it is not mentioned in the QFD literature.

(3) *Functions versus Quality Characteristics:* Functions are usually written by engineers; so this chart is often called the Voice of the Engineer versus Quality Characteristics. Purposes: to identify Functions of the product that the customer may not be aware of and to identify missing Quality Characteristics.

The Functions presented in Document 6 of Pinewood are Inspect, Impound, Racing, Judging, and Results. We made a QFD chart relating these

Functions to the Quality Characteristics. The Inspect and Impound Functions pointed out a possible new Quality Characteristic: Ensure That All Scouts Obeyed the Construction Rules.

(4) *Quality Characteristics versus Quality Characteristics:* This is the roof of the House in Figure 7.5. It is often constructed as a separate chart. Purposes: to alert the system designers to interactions, to inform the engineers whom to notify when a design change is made, and to suggest groupings of Quality Characteristics.

(5) *Quality Characteristics versus Parts:* This is the House of Figure 7.6. Purpose: to identify the Parts associated with the most important Quality Characteristics. These critical Parts might be highlighted for technological breakthroughs.

(6) *Customer Demands versus Functions:* This can also be called the Voice of the Customer versus the Voice of the Engineer. Purposes: to validate Customer Demands, to search for latent Demands that were not verbalized, to identify Functions as cost reduction targets, and to identify conflicts between the Voice of the Customer and the Voice of the Engineer.

This chart shows that the Pinewood Derby's third Performance Figure of Merit, the Happiness of the Scouts and Parents, is not explicitly addressed in the functional decomposition of Document 6 of Chapter 5. In Document 3 this task is assigned to the Grand Marshal, but the Grand Marshal is not specifically discussed in Document 6. Furthermore, this chart suggests a latent Customer Demand of Equitable Selection of Winners.

(7) *Customer Demands versus Product Failure Modes:* Purposes: to prioritize Product Failure Modes for reliability engineering and to ensure that some important Customer Demands have not been discarded.

For the Pinewood Derby, the Product Failure Modes were:

electrical failure, including
    total loss of electrical power and
    computer failure;
adverse weather conditions;
mistakes in finish line judging or in recording results;
human mistakes in
    weighing the cars,
    allowing car modifications after inspection,
    getting the cars in the correct lanes,
    lowering the starting gate,
    resetting the finish line switches, and
    wasting time; and
track imperfections that caused one lane to be faster than another.

This QFD chart must be filled out for one alternative design at a time. We first looked at the highest-ranked alternative, the Round-Robin (Best Time) tournament with Electronic Judging. This alternative is extremely sensitive to electrical failure. In fact, the race cannot be run in the event of a total electrical failure. Because of this, we designed a backup system to be used in the event of a total electrical failure: a Round Robin (Point Assignment) tournament with Human Judges. We provided the necessary charts, personnel, and training for this backup system. Figure 7.7 shows the Customer Demands versus Product Failure Modes QFD chart for the Round-Robin (Best Time) tournament with Electronic Judging with a backup system of Round Robin (Point Assignment) tournament and Human Judges.

This QFD chart shows that the most important Failure Mode for this system is mistakes in finish line judging or in recording results. This means that the systems engineer should spend the most time trying to eliminate mistakes in finish line judging or recording results. Incidentally, this provides a good definition of systems engineers: The systems engineer may be termed the "mother" of the system—the person whose job it is to worry about the

WHATs vs HOWs	Electrical failure	Total loss of electrical power	Computer failure	Adverse weather conditions	Mistakes in judging or recording	Human mistakes	Weighing cars	Allowing modifications	Correct lanes	Lowering starting gate	Resetting finish line switches	Wasting time	Track imperfections	Weight
Lots of races per scout												O		7
Very few ties		△	△		◎									5
Happy scouts		△	△	O	◎		△	△	O	△	△	O	△	10
No irate parents		△	△	O	◎		O	O	O	△	△	O	O	10
No broken cars							△			◎			△	7
Nobody touch scout's car							△		△	△				4
Fair races														
Every scout races every scout														5
Every scout races in every lane									O					5
All race about same # times														4
Don't waste time		◎	◎	O	O		△		△		O	◎	△	6
Easily implementable		◎	◎				△	△	△	△	△			9
Affordable		△	△	◎			△						△	5
Score		165	165	123	243		71	49	94	96	47	135	58	
Rank		2	2	5	1		8	10	7	6	11	4	9	

Strong relationship: ◎ 9
Medium relationship: O 3
Weak relationship: △ 1

Figure 7.7 Customer Demands versus Product Failure Modes.

success of the overall system throughout its entire life cycle. This QFD chart will help the systems engineer to focus on the most important failure modes.

(8) *Product Failure Modes versus Functions:* Purpose: to help engineers focus on the key Functions.

For the Pinewood Derby, we found that the Inspect Function was affected only by human mistakes in weighing, and the Impound Function was affected only by humans allowing modifications after inspection. These two Functions are easily dealt with and deserve little attention. The other three Functions are much more susceptible to Product Failure Modes.

(9) *Customer Demands versus New Concepts:* Purpose: to help engineers objectively review new ideas and technology. If one of the Demands of the Pinewood Derby customer was Ensure That the Scouts Did Not Place Their Cars in the Wrong Lanes, the new concept Bar Code Readers might be very enticing.

7.1.3 Advantages of using QFD

The QFD process helps detect and resolve bottlenecks. An engineering bottleneck occurs when a Quality Target Value (shown in the right column of Figure 7.6) is set at a higher level than the previous standard, at a level difficult to achieve. For the Pinewood Derby, the original standard for Judging Resolution was the 10 millisecond capability of human judges. When the target was changed to 0.1 milliseconds, a bottleneck was created. It took us three months to design a system to meet this new target value. We were fortunate that this bottleneck was discovered early in the design process. Bottlenecks discovered late in product development can cause considerable delays. Early detection of bottlenecks is a key benefit of QFD.

Japanese and American manufacturers have reported the following advantages in using QFD (King, 1989 and Akao, 1990):

- the customer needs were understood and prioritized better,
- there was increased commitment from the customer toward finalizing the design,
- design time was reduced (usually by half or a quarter),
- there was a greater awareness of the real system requirements,
- planning became more specific, which made consensus-building within the company easier,
- an informed balance between quality and cost was made,
- traceability of requirement specifications was improved,
- control points were clarified,
- the number of engineering bottlenecks was reduced,
- the design aim was communicated to manufacturing,
- there were fewer manufacturing problems at start up,
- there were fewer design changes late in development and during production,

- rework was greatly reduced,
- sales were increased,
- market share was increased,
- human relations between divisions were improved,
- employee job satisfaction was improved,
- company organization was improved, and
- the company's reputation for being serious about quality was enhanced.

However, the QFD charts obviously cannot do all of this without good systems engineering techniques.

There are three versions of every conversation (Figure 7.8 shows two of these versions): what you meant to say, what you actually said, and what the other person thought you said. As Simon and Garfunkel said in *The Boxer*, "... a man hears what he wants to hear and disregards the rest." QFD helps people communicate better; it also documents what has been agreed to.

In summary, QFD charts are rapidly becoming popular, powerful system design tools. Like the outlines shown earlier for Pinewood and SIERRA, they help ensure that important items are not overlooked. They also provide a convenient mechanism for communication between the customer and the engineer. Finally, they help to streamline the design and manufacturing process.

Figure 7.8 QFD narrows the gap between what is said and what is heard. (The Far Side cartoon by Gary Larson is reprinted by permission of Chronicle Features, San Francisco, CA.)

7.2 IDEF

IDEF is the Integrated Definition tool developed by the U.S. Air Force to model large systems. It is commonly used in the aerospace industry. The largest application of IDEF that we are aware of was the government-sponsored ECAM (Electronic Computer Aided Manufacturing) effort performed in the early 1980s, which resulted in a large collection of data related to the manufacture of a generic aerospace product. There are three parts to IDEF modeling: $IDEF_0$, $IDEF_1$, and $IDEF_2$.

$IDEF_0$ is a function modeling tool. Figure 7.9 is an example of an $IDEF_0$ chart. The information is presented to highlight what is being done at each stage, but not necessarily how it is done. Inputs are translated to outputs in a manner similar to that of state diagrams. In addition, $IDEF_0$ charts show the mechanisms used to perform the operations and the controls applied to ensure that it is done properly. Each function can be separately broken down into constituent functions, much like a system can be broken into separate subsystems. $IDEF_0$ is used much more often than $IDEF_1$ and $IDEF_2$.

$IDEF_1$ is an information modeling tool. It is an entity–attribute–relationship model. Entities possess attributes. They are related to each other through one of the following:

1. developed for/has,
2. based on/has,
3. satisfies/satisfied by, or
4 requires/required by.

The function performed and the resources used are not highlighted; instead, the emphasis is on how the information is created and related throughout the process. For example, the Product Design is *based on* the Customer Demands. The Manufacturing Process *requires* the Product Design.

$IDEF_2$ is a dynamics modeling tool. It shows work flow through a series of queues and resource allocations. The activity performed is not highlighted, rather the work flow is. This model is helpful as a time-based analysis of what is going on and when.

Many system diagramming techniques attempt to combine $IDEF_0$ with $IDEF_2$. The result is far too many functions, since each function also shows its time-based sequence in the process. The major problem with this is that many "improvements" are made by eliminating one of these functions that, in reality, must still be performed, though not necessarily at the time shown. By separating the model into two parts, it is evident that a function must still occur, but at a different time in the sequence. For example, in an engine repair operation a part may go through the degreaser twice, once when it comes out of the engine and once before it goes back in. The function Degrease is the same, but it occurs at two different times. Eliminating the second degreasing operation may reduce the cycle time of the entire process, but it will not eliminate the Degrease function with all its fixed costs. Many phony cost

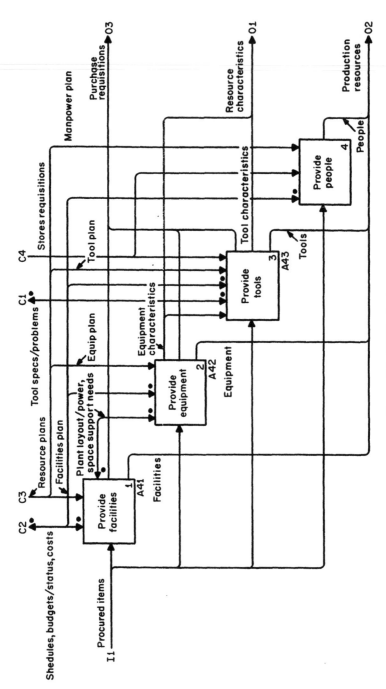

Figure 7.9 $IDEF_0$ chart.

improvements are made by eliminating operations such as the second De-grease function described above and then claiming savings in fixed costs.

7.3 Systems Engineering Design Software (SEDSO)

We decided that a software package to create system design documentation would be of great help when working with the seven systems engineering documents. The software would create the seven systems engineering documents used to track the Requirements Development and Concept Exploration phases of the life cycle. We named it Systems Engineering Design Software, or SEDSO.

SEDSO documents the system design by prompting the systems engineer for data. The questions help the engineers provide all the requirements needed in the project. Data is entered only once, though it is used in several documents. For example, the user enters data for Performance and Resource Requirements once in Document 2. This data is subsequently used in Documents 3 and 5 for evaluating the system.

SEDSO also provides templates for the design approach. For example, the sections requiring the system description are already filled in with the terminology for the system; the engineer needs only to fill in the blanks. After the state diagram is completed by the systems engineer, this data is easily entered.

Automated creation of the weights for each figure of merit is accomplished when the systems engineer enters the relative importance of each figure of merit, based on a scale of 1 to 10. These numbers are summed, and a weighted value between 0 and 1 is assigned by SEDSO as the normalized weight. Table 7.1 is an example of the weights used on SIERRA.

Sublevels of the figures of merit can also be entered. These sublevels are computed separately, and the overall value is passed up to the main level. This allows the breakdown of large figures of merit into manageable pieces. We have seldom used more than seven figures of merit at any level.

The scoring and trade-off analysis is also automated. In decision analysis many different methods are available for calculating trade-offs; we used

Table 7.1 Typical weights for the figures of merit.

Figure of Merit	Relative Importance	Weight
Number of Collisions	8	0.258
Trips by Train A	7	0.225
Trips by Train B	7	0.225
Spurious stops by A	3	0.096
Spurious stops by B	3	0.096
Availability	2	0.064
Reliability	1	0.032

scoring functions to scale the different figures of merit to values between 0 and 1. Calculation of the standard scoring function (SSF) is based on entered values for upper, lower, baseline, and slope parameters. This approach is described in Chapter 4.

The automation of the decision technique made it much easier for the systems engineer to compute the trade-offs and do the sensitivity analysis necessary for Concept Exploration in Document 5. After changing the weights, figures of merit, or parameters of the scoring function, new trade-off scores are automatically computed. These are compared with previous scores to determine how sensitive the design is to small changes in the data. For example, Table 7.2 shows approximations, or "blue sky guesses," for the number of trips completed by each train.

Using these estimates, the overall I/O Performance Index is 0.963. We now play a "what-if" game by asking what would happen if we increased the approximated values for the number of trips for Train A and Train B, as shown in Table 7.3.

The increased number of trips results in a higher scaled score, which, when multiplied by the appropriate weights and summed, yields 0.987. In other words, a design that allowed each train to make more trips had a higher overall score. This sum is the basis for computing the trade-off and determining the best system. To conduct a sensitivity analysis, the user changes the figure of merit and the software computes the new overall score. Also, the weight or scoring function could be changed for a figure of merit in Documents 2 or 3 and a new value computed in Document 5. It only takes a few minutes to change a weight, scoring function, or figure of merit and obtain a new overall score. A new document may now be output immediately with the updated information.

After the user enters all the data for a document, a copy is printed or sent to a disk file. This allows the information to be checked by others involved in the project, and it produces a permanent record of the system design. Each printed copy carries a date stamp for control purposes.

Table 7.2 Approximate values for I/O Figures of Merit.

Figure of Merit	Value	Score	Weight
Number of Collisions	0	1	0.258
Trips by Train A	8	0.917	0.225
Trips by Train B	8	0.917	0.225
Spurious stops by A	0	1	0.096
Spurious stops by B	0	1	0.096
Availability	1	1	0.064
Reliability	1	1	0.032

Overall Performance Index = 0.963

Hypertext provides a means of interlinking data that is not tied to specific fields, as it is in most flat file database schemes (Conklin, 1987). Hypertext is useful in this project, since data appears in several separate documents and because cross-checking may be done without locking users into a fixed framework. It also allows all fields in the document to be searched for a phrase. This is often needed when updating the documents.

SEDSO required a total software development time of only 180 hours, with an additional 60 hours of testing. The use of the object-oriented development tool made this possible. The size of the program is large, at 552 kilobytes, because of the integration of the database with the program code. The size of the program was acceptable considering the speed at which it was created.

Students in a graduate Systems Engineering course were asked to document a system design for a class project. The students were given the requirements for conducting a Cub Scout Pinewood Derby race similar to that described in Chapter 5. The students, working in four groups, were given the option of generating the documentation manually or with SEDSO. Three of the groups chose to use SEDSO.

The students were given one month to complete the project. This they did, averaging 200 student-hours of work and 100 pages of documentation. With the help of SEDSO as a software tool and SIERRA as an example, these students were able to write the complete set of systems engineering documents. The project that received the lowest grade was submitted by the group that chose not to use SEDSO. The instructor felt that the SEDSO-generated documents were much more consistent and had better modeling and analysis. Two of the groups generated the results with SEDSO then reformatted it using a word processor. Their documentation had the best appearance.

The students who used SEDSO felt that the software was too slow and did not have the word processing features (e.g., a spelling checker) they were accustomed to. The students found the most useful features of SEDSO to be the automatic computation of the scoring functions and the consistent transfer of requirements information from one document to the next. We felt that in

Table 7.3 Revised values for I/O Figures of Merit.

Figure of Merit	Value	Score	Weight
Number of Collisions	0	1	0.258
Trips by Train A	9	0.961	0.225
Trips by Train B	10	0.982	0.225
Spurious stops by A	0	1	0.096
Spurious stops by B	0	1	0.096
Availability	1	1	0.064
Reliability	1	1	0.032

Overall Performance Index = 0.987

completing the projects using the software the students had a much better understanding of systems theory and documentation practices. The results indicate that SEDSO was a help to the students, but a more complete and user-friendly software package is needed.

By writing our own system design documentation software, we learned a lot about what should be in a system design report. Many of the features we put into SEDSO were not even considered before the design began. Because of our hypertext environment, we decided to prompt the user to enter weights, scoring functions, test methods, test results, and sensitivity analysis for each requirement stated by the customer. This changed the way we thought about the reports, particularly during the Concept Exploration phase. We never "lost" a requirement or failed to address it, because the resulting blank paragraph gets printed in the report. The automated calculations provided us with four concept exploration analyses: present system (if one exists), approximation, simulation, and (when appropriate) prototype.

Many proposed systems are similar enough to an existing system that they can use the existing values for the figures of merit. For SIERRA, we had hundreds of previous student projects that could be used for this purpose. In the design of a mass rapid-transit system for a metropolitan area, the present system of streets and roads can be used for the figures of merit. However, many proposed systems, such as a manned station on Mars, cannot obtain figures of merit from existing systems.

We have learned from SEDSO that any systems engineering documentation aid should be able to load its own output back into its database. This became obvious when the students output their results to disk and improved the appearance of the documents with a word processor. The modified text in the word processor had to subsequently be reentered into SEDSO as well. The review and documentation process would be more efficient if SEDSO could read the document text and update the appropriate fields automatically.

Our documentation system was implemented using an IBM AT computer, but the students created their reports on an IBM XT. This computer proved to be too slow for their use, especially when they output copies of the entire set of documents.

The advantages of a computerized document system are many; the principal ones are the ability to easily keep track of requirements from one document to the next and to automatically calculate scores during concept development.

†7.4   Design for manufacturability

To design a manufacturable product means to create a product design that can be produced in a repeatable and cost-effective manner. It is easy to create

---

† This section requires some basic knowledge of probability and statistics.

a design using the most sophisticated processes money can buy and then have to throw away half the final products because they are out of specification. Unfortunately, this approach has become common among many American companies.

To design for manufacturability, the design engineer needs complete data on the parts and processes available for making the product. This "technology catalog" must consist of capability specifications, available times, costs, and quality levels.

Most designers provide nominal target values along with tolerances. For example, they may state that a hole should be located at 5.000 ± 0.003 centimeters. This implies that the designer would be happy with a hole located at 4.997 or 5.003 centimeters. However, the designer really wants all of the holes to be at 5.000 centimeters, but he is willing to tolerate a few as far away as 0.003 centimeters from the target.

The manufacturing facility also uses tolerances to describe capabilities. For example, a drill machine may have shown in the past that it is capable of drilling a hole within 0.003 centimeters of the target almost all of the time. In fact, a large amount of the time it is within 0.001 centimeters, but not always. The manufacturing engineer will say the process is capable of ±0.003 centimeter accuracy. The best way to describe what the designer wants and what manufacturing can build is with a capability curve, as shown in Figure 7.10. This curve is characterized by the mean and standard deviation of the process. Most processes can be assumed to follow a normal distribution pattern, the familiar "bell-shaped" curve. Without proving it here, it is safe to make the assumption that the capabilities of processes tend to be normally distributed. The physical characteristics of parts and processes also tend to be normally distributed. It is not, however, safe to assume that performance criteria are normally distributed. Many electronic systems output nonlinear signals that produce skewed data distributions, as shown in Figure 7.11. The top of the normal curve is the mean or average of the data, and the tails of the curve approach zero at ±3 standard deviations from the mean. The area under the curve, which represents the probability of an event occurring, is a constant at

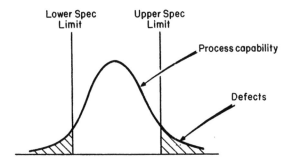

Figure 7.10 Capability curve.

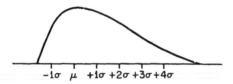

Figure 7.11   Skewed data distribution.

1.0. Figure 7.12 shows the area under the curve at various distances of the measured value from the mean. For example, the measured value is within ±1σ (standard deviations are represented by σ) from the mean 68% of the time.

We will assume that our drill machine can drill holes at 5.000 centimeters with a standard deviation of 0.001 centimeters. If the designer were to specify the hole tolerance as ±0.003 cm, the specification limits would match the capability of the process 99.7% of the time. However, this still means that 0.3% of the parts will be outside of these limits. If this hole were for a VCR that had a production run of 1,000,000 pieces, 3000 of the VCRs would be scrapped. The cost of this waste would be much too high for a cost-competitive product, such as a VCR (or any other consumer electronics product). Thus, the product is not manufacturable even though the design seems to match the capability of the manufacturing process.

Two valid options exist: either change the tolerances on the design so that the part is manufacturable or improve the manufacturing process. A design that tolerates a large variation in the process and still works is called a "robust design." Car manufacturers have improved their products recently by changing the design of the car so that the car is insensitive to variations in the manufacturing process, the environment in which it functions, and the driver.

As a rule of thumb, most products that are manufactured for mass production are designed for at least ±4σ. High quality microcircuit manufacturers in Japan design most of their products for at least ±5σ. Motorola has set its sights on Six Sigma™.

The probability of two independent parts or processes both being within tolerance can be computed by multiplying their separate within-tolerance probabilities. For example, if two holes to be drilled in the same VCR board were both specified to ±3σ, the probability that one of them would be good is 0.997, but the probability that both of them would be good is 0.997 × 0.997 = 0.994, or 99.4%. The chances that six holes would be good is $(0.997)^6 = 0.982$, or about 98%. If all the parts and processes in the product were designed for ±3σ and there were 500 parts and processes, the effective yield would be $(0.997)^{500} = 0.223$ or about 22%. This means that 78% of the products would be defective. Many tests and rework operations would have to be added to the production system to fix these defective units, and the cost of the product would skyrocket. Instead, if all of the parts and processes were designed for ±4σ, the yield would be $(0.99994)^{500} = 0.97$, or 97% yield. Still too low a yield for a mass produced item, but controllable with test and rework to allow the

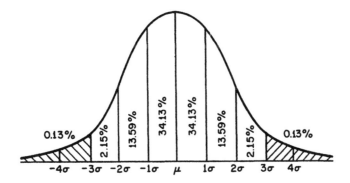

Figure 7.12 Area under a normal distribution.

product to compete in the market. At ±5σ, the product is cheap and of high quality, with a yield of $(0.9999994)^{500} = 0.9997$, or better than 99.9% good units.

7.5 *Functional decomposition*

Functional decomposition (also called top-down design) identifies the top-level functions that the system must perform and breaks them down into simpler subfunctions. Each of these subfunctions will in turn be decomposed into sub-subfunctions. This process continues until the subfunctions are finally simple enough to be specified, manufactured, and tested by a small group of people.

Research on functional analysis has shown that choosing names for the functions is important. The following suggestions may help:

1. Functions should be described by an active verb plus an object.
2. Avoid goal-like words and phrases, such as improve, reduce, and maximize.
3. Keep similar levels of abstraction in all functions.

The functions that the system must perform should be described in the Input/Output and Function Requirements. For example, the Input/Output and Function Requirements of the Pinewood Derby (discussed in Chapter 5) suggest the following top-level functions: Inspect Cars, Impound Cars, Run Races, Judge Race Outcomes, and Compute Results.

There is no unique way to do functional analysis. Sometimes it is easiest to do the physical synthesis first and let this show what the functions are. For example, the components of a radar system obviously include the radar antenna, something to move the beam, and the electronics package. These physical components suggest the following functions:

1. Send and Receive Electromagnetic Radiation,
2. Aim the Beam, and
3. Process the Signals.

Sometimes the functions can be best derived from the states of the system model. For example, in SIERRA (described in Chapter 6) the states were

1. Safe: Both trains are in the safe zone.
2. AinBout: Train A is in the danger zone, but Train B is not.
3. BinAout: Train B is in the danger zone, but Train A is not.

These states suggest the following functions:

1. Monitor Sensors to See When Either Train Enters the Danger Zone,
2. Monitor Sensors to See if Train A Leaves the Danger Zone or if Train B Attempts to Enter the Danger Zone, and
3. Monitor Sensors to See if Train B Leaves the Danger Zone or if Train A Attempts to Enter the Danger Zone.

†7.5.1 *Functional decomposition using set theoretic notation*

The mathematical statement of the Input/Output and Functional Requirements can also be functionally decomposed, as was done in Problem 4 of Chapter 5. A simpler example is given below. Let I0Rx29 be the Input/Output and Functional Requirements we are trying to satisfy.

```
IORx29 = (TRx29, IRx29, ITRx29, ORx29, OTRx29,
          MRx29)
```

where

```
TRx29 = {0, 1, 2}, /*Time interval over which the
                     Input/Output requirements
                     must be satisfied.*/
I1Rx29 = {1, 2}, /*Expected values on the first
                    input port.*/
I2Rx29 = {1, 2}, /*Expected values on the second
                    input port.*/
```

It is often convenient to manipulate the values of the inputs and outputs in these sets. When this is necessary, we will use the symbol p for the value of the input and q for the value of the output. If these variables are multiple we will append a number; here we have two input ports, so their values will be p1 and p2.

```
ITRx29 = {p1 ∈ {CNS(TRx29, 1), CNS(TRx29, 2)}},
          /*Allowed input trajectories. In this
           case, input-1 does not change in the
           time interval 0 to 2 as is indicated
           by the constant function, CNS.*/
```

---

† This section requires knowledge of our set theoretic notation. It may be skipped without loss of continuity.

```
01Rx29 = {3, 4}, /*Values that can be produced
                    on the first output.*/
02Rx29 = {3, 4}, /*Values that can be produced
                    on the second output.*/
OTRx29 = {q2 ∈ {CNS(TRx29, 3), CNS(TRx29, 4)}},
            /*Allowed output trajectories. In this
              case, output-2 is a constant over the
              given time interval.*/
MRx29 = {if (p2(0) = 1)         .
            then q2 = 3; else q2 = 4}.
          /*The Input/Output matching function
            that must be satisfied. In this case,
            if input-2 at time zero is a 1, then
            output-2 should be a 3 for all times,
            otherwise it should be a 4.*/
```

Suppose that last year your company built systems $Zx25$ and $Zx26$ that satisfied Input/Output and Functional Requirements $IORx25$ and $IORx26$. If we can prove that $IORx29$ decomposes into $IORx25$ and $IORx26$, we will not have to design a new system.

```
IORx25 = (TRx25, IRx25, ITRx25, ORx25, OTRx25,
          MRx25)
```

where

```
TRx25 = {0, 1, 2},
I1Rx25 = {5, 6},
I2Rx25 = {1, 2},
ITRx25 = {p2 ∈ {CNS(TRx25+, 1), CNS(TRx25+, 2)}},
            /*Input-2 must not change in the time
              interval 0 to 2.*/
01Rx25 = {3, 4},
02Rx25 = {7, 8},
OTRx25 = {q2 ∈ {CNS(TRx25+, 7), CNS(TRx25+, 8)}},
            /*Output-2 is a constant with a value
              of either 7 or 8.*/
MRx25 = {if (p2(0) = 1)
            then q2 = 7; else q2 = 8}.
          /*If input-2 at time zero is a 1 then
            output-2 should be a 7 for all time,
 ⏐        otherwise it should be an 8.*/

IORx26 = (TRx26, IRx26, ITRx26, ORx26, OTRx26,
          MRx26)
```

where

```
TRx26 = {0, 1, 2},
```

```
I1Rx26 = {7, 8},
I2Rx26 = {1, 2},
ITRx26 = FNS(TRx26, IRx26),
         /*All possible functions that can be
            made from TRx26 and IRx26; i.e.,
            there are no input trajectory
            restrictions.*/
01Rx26 = {5, 6},
02Rx26 = {3, 4},
OTRx26 = {q1 ∈ {CNS(TRx26+, 5), CNS(TRx26+, 6)};
          AND q2 ∈ {CNS(TRx26+, 3),
                         CNS(TRx26+, 4)}},
         /*Both outputs must be constants.*/
MRx26 = {if (p1(0) = 7)
            then q1 = 5; else q1 = 6;
            AND if (p2(0) = 1)
            then q2 = 3; else q2 = 4}.
         /*If input-1 at time zero is a 7 then
            output-1 should be a 5 for all time,
            otherwise it should be a 6. If input-2
            at time zero is a 1 then output-2
            should be a 3 for all time, otherwise
            it should be a 4.*/
```

Figure 7.13 shows that I0Rx25 and I0Rx26 can indeed be coupled to yield the desired behavior of I0Rx29. The input I2Rx29 was not restricted, and Figure 7.13 shows that it can accept any trajectories of 1's and 2's. The output 01Rx29 was never used, and Figure 7.13 shows that this output will produce unrestricted trajectories of 3's and 4's. Therefore the input–output behavior of the coupled requirements is the same as the desired input–output behavior. We have reduced our problem to a previously solved problem.

Once upon a time, an engineering professor and a mathematics professor were bragging about the quality of their students. They decided to have a contest to determine who had the best students. Each professor selected one representative student and brought him to the demonstration. First, the engineering student was brought into a classroom filled with spectators. He was turned to face the audience, then someone threw a match into the wastepaper basket behind him. The paper burst into flame. The engineering student looked around the room frantically and saw a bucket of water on the desk in front of him. He quickly dumped it on the flaming wastebasket and put out the fire. The engineering student, buoyed by the applause of the audience, was then escorted from the room. Next, the mathematics student was brought into the classroom and he was turned to face the audience. Once again, someone behind him threw a match into the wastepaper basket. The paper burst into flame. The mathematics student looked around the room frantically

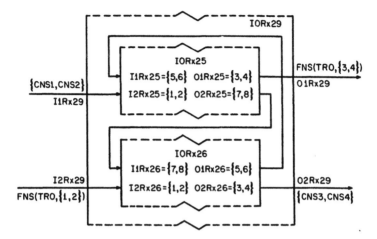

Figure 7.13 Functional decomposition of Input/Output and Functional Requirements.

and saw a bucket of water on the desk in front of him. He quickly dumped it on the flaming wastebasket and put out the fire. The mathematics student, buoyed by the applause of the audience, was then escorted from the room.

The second phase of the contest now began. First, the engineering student was brought into the classroom. He was turned to face the audience, and someone behind him threw a match into a wastepaper basket. The paper burst into flame. The engineering student looked at the table in front of him, but the bucket of water was not there. He furiously scoured the whole room and finally saw a bucket of water on the floor way in the back of the room. He rushed to the back, grabbed the bucket, rushed back to the front of the room, and doused the fire. The audience applauded, and the highly acclaimed engineering student was escorted from the room. The mathematics student was then returned to the room. He was turned to face the audience and someone behind him threw a match into the wastepaper basket. The paper burst into flame. The mathematics student looked at the table in front of him, but the bucket of water was not there. He furiously scoured the whole room and finally saw a bucket of water on the floor way in the back of the room. He rushed to the back, grabbed the bucket, rushed back to the front of the room, and put the bucket on the table in front of him. The audience was crestfallen. After a pregnant pause, the mathematics professor exclaimed, "What did you do?" The mathematics student replied, "I have reduced this problem to a previously solved problem."

7.6 *Concurrent engineering*

Several years ago the chief executive officer of a major Japanese company said, "In United States manufacturing you have all-star teams, but you keep losing all the games." Why is this? If the winner of football's Super Bowl were to play a team of all-stars, the Super Bowl winner would probably win. Why? Because of a better game plan and better team work. That is what concurrent engineering tries to provide in the manufacturing field (Clausing, 1990).

Lake (1991) reported two government efforts to define concurrent engineering. (1) An Institute for Defense Analysis report (IDA Report R-388) defined concurrent engineering as "a systematic approach to the integrated, concurrent design of products and their related processes, including manufacturing and support. This approach is intended to cause the developers, from the onset, to consider all elements of the product life cycle from conception through disposal, including quality, cost, schedule and user requirements." (2) The U.S. Army Communications–Electronics Command has a Concurrent Engineering Directorate who say, "Concurrent Engineering is the simultaneous and integrated engineering of all design, manufacturing, and support aspects of a product from concept through availability. It is a teaming concept. All the people who normally get involved in the product come together as a team. They work together, trading ideas and ensuring what they decide now (like design decisions, or major product modifications) will not adversely affect what they have to do later (like manufacture in quality, or ensure support in the field). Everything is addressed simultaneously."

How do you do concurrent engineering? If you have read this book carefully, you know how. Systems engineering, as presented in this book, *is* concurrent engineering. It seems to be the consensus among practitioners that concurrent engineering is systems engineering as it should be done, not as it is actually practiced. One of the important concepts of concurrent engineering is the use of multi-disciplinary teams, which was articulated for systems engineering long ago by Wymore (1976).

Traditional design and manufacturing engineering has been characterized as the "throw it over the wall" technique. That is, the marketing department determines the customer's needs and throws them over the wall to the planners, who outline the requirements for the product and throw them over the wall to the engineers, who design the system and throw their plans over the wall to the manufacturing department, and so on. This is also called a "stove-pipe" organizational structure: a manager talks to employees in her division, to her boss and to her boss's boss, but not to people in other divisions; the design engineers only talk to other design engineers, the manufacturing engineers only talk to other manufacturing engineers, and the quality control people only talk to other quality control people.

In concurrent engineering, the product is designed concurrent with the development of production capability, field support, and quality engineering. We do not design a product—we design a system that produces a product. Clausing (1990) describes the benefits to the product of this concurrency:

1. field support personnel and quality engineers can have an earlier start,

2. trade-offs can be made involving design, production, and maintainability issues, not simply within the design,

3. designing for manufacturability and designing for maintainability are facilitated,

4. production and field support personnel understand the design and are committed to its support, and

5. prototype iterations are reduced.

In conventional engineering, an electrical engineer might struggle to minimize power usage or maximize the signal-to-noise ratio, but in concurrent engineering, electrical performance is no longer of paramount importance. Rather, the goal is to get the best design in terms of total quality, which means achieving a balance in performance, ease of manufacture, and ease of support. The early exchange of information is inherent in concurrent engineering; this allows manufacturing personnel to know which dimensions are critical to the performance of the product and which can be relaxed to improve manufacturing efficiency and yield (*IEEE Spectrum*, 1991).

The concurrent engineering process provides these benefits to management:

1. Consideration of all players and all phases of the system life cycle produces a better design process.

2. It allows management to concentrate on quality, cost, and schedule.

3. The voice of the customer is emphasized, and internal company metrics are not emphasized.

4. For all processes, alternatives are investigated and trade studies are performed to find the best design. The alternatives must include the competition's products. This process is also called competitive benchmarking.

5. The people in the company become closer knit, working together and understanding each other better. Communications occur horizontally and are not overly constrained by the vertical orientation of tree-like organization charts.

6. The same multi-disciplinary team follows the product from start to finish, enabling them to better understand the product.

7. Suppliers become an integral part of the design team. In traditional engineering, the suppliers are brought in at a very late stage. They are told, "Here is what we need; at what price can you supply it?" This leads to a proliferation of suppliers, none of whom contribute to the design process. In concurrent engineering, the supplier base is reduced and the participating suppliers are treated as members of the design team. At Xerox, implementing concurrent engineering reduced the number of suppliers from over 3000 to under 400 (Clausing, 1990).

To carry this last train of thought one step further, explaining your needs to your supplier's supplier might reduce your supplier's cost, and consequently your own. On the other side of the issue, listening to your customer's customer might allow you to manufacture a product that is more satisfactory to the ultimate customer. If you are very good at communicating with your supplier's supplier and your customer's customer, you may even be able to reduce your work to nothing, though this may not be good for your business.

The following example shows how talking to your customer's customer can be beneficial. Assume that you work for a medical supplies distributor and that you sell, among other things, hypodermic needles to physicians. When you talk to the nurses and patients of a medical practice, you find out that they prefer sharp needles. You then talk to your supplier's supplier, the company that makes the machines that make the hypodermic needles. You discover that they have a new laser technique for making microelectrodes, which are used to record the electrical activity of single neurons in animals, and that this laser technique can be modified to make super-sharp hypodermic needles. If you can convince your supplier to use the new machines, your efforts would produce substantial added value to the product. However, talking to your customer's customer and your supplier's supplier will not always be possible. Bureaucratic, geographical, or communications barriers often prevent it, but it is worth a try.

Since it is unusual to get something for nothing, there must be additional costs involved with concurrent engineering.

- A product's start-to-finish time is usually shorter with concurrent engineering, so the concurrent engineering process must require more people. If more people are to understand more about the product, more time must be spent to teach them.

- There are more up-front costs, so there must be a strong initial commitment to the product.

- The size of the multi-disciplinary design team must change many times in the design process, which may cause difficulties for many companies.

- Everyone involved must be educated about the new process and their role in it. For example, in the early stages of the system design, manufacturing people often bog down the process with premature discussions of nuts and bolts issues. They must be taught that the system design process is iterative. The overall system is first designed and then broken down into subsystems or components. The system design process is applied to these subsystems. Nuts and bolts issues must be saved for design meetings about the subsystems.

A series of government and industry workshops have identified the following ten impediments to effective implementation of concurrent engineering (Lake, 1991).

1. Few people currently understand concurrent engineering.

2. The serial approach to system development has been institutionalized.

3. No culture currently supports concurrent engineering.

4. Organizational roles are not defined. .

5. There are no incentives to induce a contractor to switch to concurrent engineering.

6. Up-front resources to support concurrent engineering are difficult to obtain.

7. Currently there is little discipline in the requirements definition process.

8. Current contracting techniques do not support concurrent engineering.

9. Currently, risk considerations are not done up front.

10. The government dictates design and business practices, inhibiting the contractor's ability to use innovative approaches.

To help solve these problems, the workshops suggested, among other things, that university engineering curricula include use of the following specialized tools:

- quality function deployment (QFD),
- the design of experiments,
- statistical process control,
- CAD/CAM/CAE/CIM,
- Ishikawa (cause–effect fishbone) diagrams,
- Taguchi methods,
- affinity diagrams,
- computer-aided system engineering,
- common design data base,
- computer-aided software development (CASE), and
- computer-aided logistics support.

They emphasize that these tools should then be used in multi-disciplinary project courses.

7.7   *New trends in engineering design*

Some of the themes expressed in this book represent new trends in the philosophy of engineering design, such as:

1. flexibility being more important than efficiency,

2. dealing with uncertainty being more important than precision,

3. the voice of the customer being more important than the voice of the engineer, and

4. problem stating being more important than problem solving.

The lunar lander described in Chapter 2 had a flexible design and was very successful, but it was not efficient because its capabilities exceeded the requirements. The Hubble telescope, also discussed in Chapter 2, required precision manufacturing and could not cope with uncertain measurements; it was not as successful.

In some areas of human endeavor, such as professional sports, efficiency is more important than flexibility. Few professional baseball players play both pitcher and catcher positions. They specialize in one position or the other to become more efficient players. However, in modern engineering design, flexibility is becoming more important than efficiency. In some old engineering curricula, optimization and operations research were the most important tools. Recently, multi-objective decision making, managing uncertainty, the voice of the customer, fuzzy logic, total quality management, and design for manufacturability have become more important.

Personal computers (PCs) are very flexible tools. They are used by some people for text processing and by others for accounting. The PC can be used for signal processing or process control with the addition of analog-to-digital and digital-to-analog converters. It can be used to simulate artificial neural networks by adding parallel processing boards. Special purpose machines that do each of these tasks are available, but most people use the flexible PC. If the PC design engineers were to increase the number of instructions that the PC could handle, it would become more efficient for many tasks. But PCs were designed for flexibility, not efficiency. Another example of flexibility is the Motorola 68000 microprocessor, which is used in Apple's Macintosh computer, as well as in computers by Bull, Commodore, Fujitsu, GMX, Hewlett-Packard, Motorola, NEC, NeXT, Texas Instruments, and others. It is also used in many laser printers and other intelligent devices. It is flexible enough to be used by many companies for many products.

Part of the reason that flexibility is becoming more important than efficiency is the rapid rate at which technology is advancing. For example, a company could buy a certain device, spend a year optimizing its inclusion into their system, and then discover that a new device is available that performs better than their customized system and at less cost. In the days of vacuum tube radios, every few years an engineer would design a circuit to eliminate one vacuum tube; he would name the circuit after himself and apply for a patent. With modern VLSI circuits, saving even a few thousand transistors is not an impressive accomplishment—the Motorola 68040 microprocessor has over a million transistors.

Reusability is a subset of flexibility. The best examples come from computer software. In the early 1970s, various computer users running the Unix operating system wrote hundreds of small stand-alone programs like the following: *tr* transliterates a file, *sort* merges input files together, *uniq* compares adjacent lines in a file, *comm* reads two input files and marks lines that are common to both, and *deroff* removes all formatting commands from a text file. Each of these programs was written for a specific purpose, but they can be reused for something different. The following line produces an index for a book:

deroff -w filename | tr A-Z a-z | sort | uniq | comm -23 - common index

The phrase "deroff -w filename" removes all formatting commands from the file called filename, and the -w option arranges the file with one word per line. This output then becomes the input for "tr A-Z a-z," which changes upper-case letters into lower-case letters. The output of this process is passed to "sort," which alphabetizes the file. The program "uniq" then removes all duplicate lines. Next, the section "comm -23 - common" produces an output containing all those words in the input file that are not in the list of the most common English words. This final output file is the start of an index for a book, although creating an index for a book never entered the minds of the designers of these programs. By putting in a little bit of thought to the design process they were able to make programs that could be used by others for a broad range of purposes. Admittedly, most present word processors have specialized built-in indexing programs. They are fancier, and they certainly cost a lot more. But these programs do not have the flexibility to be used for other purposes. They are also limited to indexing words instead of concepts.

Of course, the number of items to be manufactured affects the flexibility–efficiency trade-off. For three custom-made communications satellites, engineers will try to reuse as much as possible from previous projects—the efficiency of the system will not be important. However, if your company hopes to sell one million word processors, it may be willing to put in a lot of time optimizing the product.

At one time, engineering students were taught to build precise systems. However, there is a lot of uncertainty in the modern world, and precise systems can seldom deal with uncertain circumstances. In designing a system for shooting clay pigeons, you could specify a Remington®700 Mountain Rifle in 7mm Mauser with an Aimpoint® 5000 electronic sight and hand-loaded 162-grain Nosler solid-base bullets, but a shotgun would probably be better able to cope with the uncertainties in the disk's flight, the variable wind currents, and the hunter's neuromuscular system. Often, we cannot even state a precise input–output specification; we can only bound the behavior, perhaps using matching functions to show the output behaviors that are acceptable for particular inputs, as was shown in the windshield wiper system of Problem 3 in Chapter 4.

In Section 7.4 we discussed designing for manufacturability. With that in mind, consider the simplified process of drilling two holes in a board and attaching a handle. We could buy handles with two flat-bottomed pegs, ¼ inch diameter, spaced five inches apart that require holes 0.25 ± 0.0001 inches in diameter be drilled in the boards 5.0 ± 0.0001 inches apart. Alternatively, we could reduce the required precision of the manufacturing process and increase our chances of success by buying flexible handles with round-bottom pegs and by specifying hole diameters of 0.28 ± 0.01 inches spaced 5.0 ± 0.01 inches apart. The later design would still function despite variances in the environment or in the manufacturing process. The resulting system is termed robust. Systems should be designed whenever possible so that high precision is not needed and so that they will work despite unexpected variances.

If an engineer designs the world's most precise and efficient widget, but customers do not buy it, the engineer is a failure. The engineer must listen to the voice of the customer. As exemplified by Sections 5.1.3 and 6.1.3 (in Chapters 5 and 6, respectively), the "customer" may actually consist of a disparate group that must be satisfied for a successful design. The customer's "voice" amounts to the collection of demands that require accommodation, which is discussed in Section 7.1 (in this chapter).

Finally, as is stated so often in this book, problem solving is not as important as problem stating.

Problems

1. **Engineering education.**   A university has many customers, but if we concentrate only on undergraduate engineering education at state schools, we can narrow them down to students, parents, the state government, taxpayers, and industry. Let us ignore the first four customers and only consider industry. We asked several industry leaders *what* characteristics they felt were important in new engineering graduates. They said a new engineering graduate should Have a Basic Engineering Background, Be Ready to Participate in Design Projects, Have Interdisciplinary Knowledge, Be Computer Literate, Be Aware of Manufacturing Processes, Be Able to Write and Speak Well, and Be Able to Solve Small Practical Problems. We then asked industry and university leaders to name the measures that might be available to see *how* well these customer wants were met. They suggested: Whether the Program Was Accredited, the Number of Hours of Laboratory Work, the Number of Large Projects the Students Participated in, the Hours of Computer Usage, the Use of Manufacturing Processes, and the Number of Courses Taken That Had a Writing or Speaking Emphasis. Construct a quality function deployment (QFD) chart, like the one shown in Figure 7.4, for these Whats and Hows. Provide whatever weights and relationship values you think are best. By your calculations, what are the most important measures that an employer should look for?

Next, we asked the university administrators what instruments they had available for implementing these measures. They said that from the ABET accreditation materials they could readily provide the number of units in the curriculum described by Mathematics and Basic Science, Engineering Science, Engineering Design, and the Humanities and Social Sciences. With just a little bit more effort they could provide the number of units in the curriculum described by Engineering Laboratories, Engineering Project Courses, Science Laboratories, and Courses with a Writing Emphasis. Construct a QFD

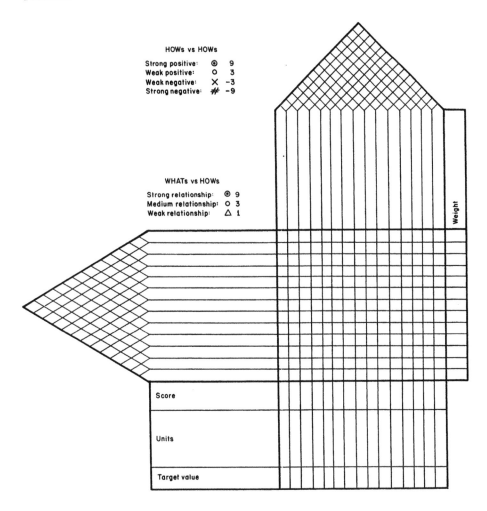

chart relating these measures and instruments. On which of these instruments should the university administrators concentrate in order to keep their industrial customers happy?

2. **Teaching a course.** Assume that you want to teach a systems engineering course in the coming semester. Your customers will be your students and your department head. Construct a quality function deployment (QFD) chart, like the one shown in Figure 7.4, relating your customer's demands (the Whats) and the measures you will use to see if your customer's demands are satisfied (the Hows). Which of your measures is most important? That is, which received the highest score?

3. **Context.** Context is very important in evaluating statements. For example, suppose someone said to you, "I like your shirt." You might interpret

that as being complimentary. However, suppose that the context of that statement is different—the comment was in reply to your question, "Well, is there anything about me that you *do* like?" Perhaps, in this instance, you would not feel so flattered.

4. **Statistics abuse.** Three men were out hunting and a rabbit dashed across their path. The first man shot six inches in front of the rabbit. The second man shot six inches behind the rabbit. The third man, a statistician, exclaimed, "We got him!"

chapter eight

Projects

The best way to learn how to design systems is to design systems: students should do system design projects. At the beginning of the semester, the projects can be small, taking no more than 10 student-hours, but at the end of the semester, a project requiring 100 student-hours is appropriate. It is important that these projects be done by groups of two to five students, because in the real world, systems are designed by teams. In this chapter we present three types of projects: (1) simple projects that require little knowledge of "physics," (2) simple projects that require specific knowledge about the problem domain, and (3) large projects.

The instructor should specify the documents to be written for each project. It is worthwhile to write all seven systems engineering documents for a few projects, but for most projects only parts of Documents 2, 3, and 5 should be written, since writing all seven documents is a very time-consuming task. The most important tasks are specifying the requirements in Documents 2 and 3 and performing the trade-off study of Document 5. In the real world, Documents 4, 6, and 7 would only be written for big projects. Students should keep track of the time and money invested on projects in which the design is followed through its entire life cycle, and they should compare this data with Figure 2.2.

For some of these projects the figures of merit are specified, for others they are only suggested. Choosing figures of merit, assigning weights to the figures of merit, and writing an equation to combine the figures of merit into the final trade-off figure of merit are important tasks that must be done by the class and instructor before the projects are begun.

Students often expect projects to satisfy the reciprocal principle. This principle states: "All the data needed is provided and all the data provided is needed" (Clements, 1989). An extension to this principle says: "The accuracy of the provided data is appropriate to the problem." Unfortunately, this principle seldom applies in the real world. A major part in analyzing these projects will be finding the needed data and distinguishing between relevant and irrelevant data.

8.1 Regular projects

1. **Home security systems.** Split the class into groups of three to five students. Pick one student in each group to be the customer. Design a home security system for that person. Write an Operational Need Document and do a trade-off study of your alternatives. Systems engineering methodology is more important here than complete accuracy. Study your textbook, not home security brochures. For this exercise, you already know everything you need to know about home security systems! In your trade-off study be sure to compare your alternatives to our favorites: (1) Do nothing. (2) Adopt two German shepherd dogs.

Be sure to do a "what-if" analysis of your system. For example, if you are using dogs, explain how the system will behave if one dog dies. If, in response to a break-in, your proposed system automatically telephones a security office that contacts the police, explain how your system performs if the burglar first cuts your telephone line.

2. **The U-tube accelerometer.** An inventor has just come to you with the device shown in Figure 8.1. She claims that it will measure acceleration, which could be useful in many applications, such as activating seat belts and air bags in automobiles. She says that in order to get financial backing she has to have a set of documents that describe her system. Prepare the seven systems engineering documents shown in this book for this system.

This is her explanation of how the device works: The device is a glass tube bent into a rectangular U shape and filled with colored liquid. It is to be mounted in the automobile with the bottom part of the tube oriented fore and aft and with the sides vertical. When the car moves at a constant velocity, the liquid in the two vertical arms has the same height; if the car accelerates, the liquid rises in the back tube and falls in the front tube; if the brakes are applied, decelerating the car, the liquid rises in the front tube and falls in the back tube. The change in level in one vertical tube, as read off the scale, gives the amount

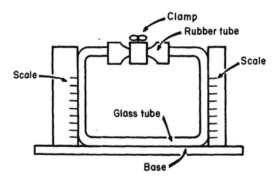

Figure 8.1 The U-tube accelerometer.

of acceleration or deceleration. Thus, the proposed instrument could be used to test the ability of a vehicle to accelerate or the capacity of its brakes to stop it. It could also be used to activate seat belts and air bags. The tops of the U-tubes are connected with a rubber tube and clamp to impede the flow when the levels are changing. This damps out oscillations that occur when the acceleration changes suddenly. This project is based on Ver Planck and Teare (1954).

As an optional part of this project assignment, evaluate the following enhancements: The inventor believes that making the bottom horizontal tube larger than the vertical ones makes the instrument more sensitive; that is, it gives a greater change of level for a given acceleration. She also proposes filling the upper portion of the tube (above the colored liquid) with a fluid of lighter density than the primary fluid. Do you think either of these changes will prove useful? As a second option, investigate the effect on the system of going up and down hills. That is, compare the sensitivity of the device during rapid decelerations to its sensitivity when going down a steep hill. As a third option, build such a U-tube accelerometer and test the effectiveness of the damping clamp. As a fourth option, discuss the consequences of the tube breaking.

3. **Baseball.** Model the flight of an 80 mph curve ball between the pitcher's hand and the batter's bat. Discuss how you could test how good your model was. The following is a simple explanation of the physics of the curve ball (see Watts and Bahill, 1990, for details).

There is no longer a controversy about whether or not a curve ball curves. It does. The curve ball obeys the laws of physics. These laws say that the spin of the ball causes the curve. If the spin is horizontal (as on a toy top), the ball curves horizontally. If it is top spin, the ball drops more than it would due to gravity alone. If the spin is somewhere in between these two planes, the ball both curves and drops. In baseball, most curve balls curve horizontally and drop vertically. The advantage of the drop is that the "sweet spot"(the best hitting area) on the bat is about six inches long but only one-half inch high. Thus, a vertical drop is more effective at taking the ball away from the bat's sweet spot than a horizontal curve. We now present the principles of physics that explain why the curve ball curves.

The first part of our explanation invokes Bernoulli's principle. When a spinning ball is placed in moving air, as shown in Figure 8.2, the movement of the surface of the ball and the thin layer of air that "sticks" to it slows down the air flowing over the top of the ball and speeds up the air flowing underneath the ball. According to Bernoulli's equation, the area with the lower speed (the top) has the higher pressure, and the area with the higher speed (the bottom) has the lower pressure. This difference in pressure pushes the ball downward. If this were considered from the perspective of a top view, it would explain the horizontal curve of a curve ball. If this were considered from the perspective of a side view, it would explain the abrupt drop of the ball.

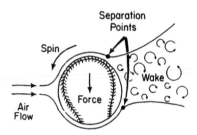

Figure 8.2 Air flow around a spinning baseball.

The second, and probably more important, part of our explanation involves the wake of chaotic air behind the ball. Air flows smoothly around the ball until it gets to the separation points, where it changes into a chaotic, swirling flow called the wake. It is easy to see a wake in water on the downstream side of bridge supports and behind boats. In a boat, swinging the rudder to the right deflects water to the right and, according to the principle of conservation of momentum, the back of the boat must be pushed to the left. You can feel this force if you put your hand out of the window of a moving car. (Make sure the driver knows you are doing this!) Hold your hand horizontal with your palm down. Now tilt the front of your palm upward so that the wind hits the palm of your hand at an angle. This deflects the air downward, which causes your hand to be pushed upward. We now relate this to the spinning baseball in Figure 8.2. Before the ball interacts with the air, all the momentum is horizontal. The wake that develops in the air due to the ball's spin has upward momentum. The principle of conservation of momentum requires, therefore, that the ball have downward momentum, meaning that the ball will be pushed downward.

There are several ways to shift the wake behind a baseball. The wake is shifted by the spin on a curve ball. The friction that slows down the flow of air over the top of the ball causes the air to separate from the ball sooner on the top than on the bottom, as shown in Figure 8.2. This shifts the wake upward and pushes the ball downward. For nonspinning pitches, such as the knuckle ball and the scuff ball, when the seams or the scuff are near the bottom separation point they create turbulence, which delays the separation, as shown at the bottom of the figure. This again shifts the wake upward and pushes the ball downward.

When the pitcher puts horizontal spin on the ball, the wake of chaotic air behind the ball is moved to one side, causing the ball to curve and thereby confounding the poor batter who is trying to hit it.

The lateral force on the ball is given quantitatively by

$$F_{\text{lateral}} = \tfrac{1}{2}\rho\pi R^3 \omega v$$

where ρ is the air density (0.0023 is a common value), R is the radius of the ball (0.119 feet), ω is its spin rate (190 radians per second is reasonable for a curve ball), and v is its velocity (in feet per second). The direction of this force will be horizontal, vertical, or something in between, depending on the axis of rotation. The ball will also decelerate due to the drag force of the air:

$$F_{drag} = \tfrac{1}{2}\rho\pi R^2 C_D v^2$$

where ρ is the air density, R is the radius of the ball, C_D is the coefficient of drag (with a value of about 0.5), and v is the ball's velocity.

Most systems are impossible to model in their entirety, but they are composed of hierarchies of subsystems that can be modeled. Simon (1962) discusses the need for such hierarchies in complex systems. He shows that most complex systems are decomposable, enabling subsystems to be modeled outside the entire hierarchy. For example, in studying the motion of a baseball it is sufficient to apply Newtonian mechanics considering only gravity, air, the ball, and the bat. One need not worry about electron orbits or the motions of the sun and moon. Forces that are important when studying objects of one order of magnitude seldom have an effect on objects of another order of magnitude. This is an important principle of modeling and is amply demonstrated in this project.

4. **Successive refinements of the model for celestial mechanics.** Ptolemy (who lived in the second century A.D.) wrote that the sun revolves around the earth. Copernicus (1473–1543) postulated that the earth circles the sun. Galileo (1564–1642) gathered observational data that confirmed the Copernican model. Kepler (1571–1630) demonstrated that the earth's orbit is elliptical, with the sun at one focus. Newton's (1642–1727) law of gravitation explained that the sun and the earth rotate about their center of mass and showed how other planets and moons affect this trajectory. Einstein's (1879–1955) theory of relativity handled the observed anomaly in Mercury's orbit (for an interesting discussion of this, see Jeffreys and Berger, 1992) and went on to make two more testable predictions: that gravity bends light rays and that clocks slow down in a gravitational field.

Do you think that Einstein's theory will ever be proven wrong? Describe some forces, objects, relationships, or structures that were not known in the early part of the twentieth century that might cause a refinement of Einstein's theory of relativity. Develop another example of model refinement throughout history.

5. **Modeling a comet.** Comets have been modeled as dirty snowballs. To help refine this model, scientists need to obtain temperature and density profiles from the surface to the core of a comet. Design a system to obtain this data. Although this is not a restriction on the technology you may use, most of your solutions will probably involve launching a probe from a spacecraft. For spacecraft-based solutions, your probe must not exceed one kilogram in mass, 2.5 centimeters in diameter, and 120 centimeters in length. (This project was suggested by Ray Buza of Martin Marietta, Denver, CO.)

6. **Diameter of the earth.** Design a system to measure the diameter of the earth. Be sure to specify your accuracy.

7. **Boat bailers.** Boating has become a national pastime; there are over a million small boats on the rivers, lakes, and seacoasts of this country. For the most part, they belong to weekend sailors. Therefore, the boats sit at their moorings for weeks at a time unattended.

A common problem with such craft is minor leaking through propeller shaft stuffing boxes, centerboard trunks, etc. It is too expensive to hire a watchman, and battery-powered bilge pumps would run down most batteries. Boat owners want a simple bailer that operates from the natural energy found in the environment (solar power, wind, water current, or wave action). Note that the rocking of the boat as a result of wave action develops considerable kinetic energy.

Your task, should you choose to accept it, is to design a system to remove water from a boat. Your system should be inexpensive to build, easy to install, and require maintenance less than than once a month, and it should not get in the way when the boat is being used. Your system must also have sales appeal. For rough design purposes, assume a leakage rate of one or two liters per day and assume that the boat rocks ±5 degrees ten times per minute.

We have thought of several possible solutions and some problems with each of these solutions. A piston and cylinder attached to the mooring line would work well, except that if the cylinder is long and the pump piston leaks, the pump must be primed. A round ball in a cylinder on the bottom of the boat would make a good solution, but unfortunately good one-way valves are hard to make. A hamster in a squirrel cage would work well, but animals require automatic feeding machines or daily attention.

8. **Water for California.** Southern California often has water shortages. Design a system to help the city of San Diego solve its water problem. Some solutions that have already been suggested are desalinating seawater (it costs $30 million to build a typical plant) and towing an iceberg from the Arctic Ocean. One difficulty associated with the latter solution is attaching the cable

Typical California water rates
(in dollars per acre-foot[†])

Household water	300
Agricultural water	100
Reclaimed sewage	450
Tanker ships from Vancouver	1500
Existing desalination plants	1000
New desalination plants	2000

[†]One acre-foot is 325,852 gallons, or enough to cover one acre with one foot of water.

to the iceberg, because pressure causes the ice to melt. If our table is helpful, use it. But you should attempt to validate it, because the price of water is determined by political, not economic, considerations. For comparison, China's Guangdong Province sells water to Hong Kong for $240 per acre-foot.

Other alternatives under consideration include an undersea pipeline from Alaska that would cost $150 billion and deliver 12 trillion gallons of water per year and an overland pipeline from British Columbia's North Thompson River that would cost $3.8 billion.

9. **"Weighing" an astronaut.** Physicians working for NASA want a system to monitor the mass of astronauts when they are weightless on extended missions, such as in a space station or on a trip to Mars. Imagine that your company is making a proposal to NASA to develop such a system. Write the systems engineering documents for your proposed system.

Think of physical principles that involve mass, but not gravity, such as Newton's second law, the conservation of momentum, the period of simple harmonic motion, etc. Important criteria for making a final choice of method are that the device have low mass and a small size. Cost is not important. The physicians would like to know the mass to within 100 grams, but they may settle for a resolution of a kilogram. One difficulty you may encounter is that human bodies are not rigid; arms, legs, and even internal organs tend to move under acceleration, and acceleration may well be a part of the method you consider.

Be sure to describe how you will test your system to ensure that it works. Remember that though it will be designed to work in space, you will presumably test it on earth first. In addition, design an on-board system to test your device so that we know that it is functioning properly after it is launched. You could propose sending up an 80 kilogram bag of water to be used as a test load, but it is very expensive to launch extra mass into orbit.

10. **The Popsicle® stick bridge.** This project should be preceded by a lecture or two about engineering statics. Using Popsicle sticks, paper clips, and Elmer's® white glue, build a bridge truss system that will span a 25 centimeter chasm and support a centrally applied vertical force of 89 newtons. The best design will be judged to be the one that resists the greatest force before breaking and uses the fewest paper clips, Popsicle sticks, and glue.

Your system will be tested on two tables 25 centimeters apart. A vertical force of 89 newtons will be slowly applied by means of a string and one inch diameter washer. The center of your system must accommodate this string and washer. You may not tie knots in the string. Once the force reaches 89 newtons, it will remain for ten seconds. Then the force will be slowly increased until the structure breaks. The breaking force will be recorded and will be used to help calculate the overall Performance Figure of Merit.

The overall performance figure of merit (IF0P) will be based on the following performance figures of merit:

1. Weight Supported Before Failure and
2. Appearance.

The overall Utilization of Resources Figure of Merit (UF0P) will be based on the following resource figures of merit:

1. Number of Popsicle Sticks,
2. Number of Paper Clips,
3. Amount of Glue, and
4. Total Weight of the Bridge.

The final Trade-Off Figure of Merit will be defined as

$$\text{TF0P} = \frac{\text{IF0P}}{\text{UF0P}}$$

Before construction begins each class should decide on the weights for these figures of merit. In the beginning of your design process you may want to read R. S. Kirby et al., *Engineering in History*, McGraw-Hill, New York, 1956, pp. 220–244.

11. **Traffic synchronization.** Design a system for synchronizing the traffic signals on a section of a busy street in your town. A synchronized system is one in which the green lights cascade in each direction, enabling a vehicle to maintain a cruising speed of 40 kilometers per hour without stopping. We want to know how long it would take for the cost of the energy saved by the vehicles to pay off the cost of installation and maintenance. The class should be split into groups of two to five students. Define the problem, plan your solution (including data collection and data analysis), and write the systems engineering documents. This project should be preceded by a lecture and a homework set on the time value of money.

12. **Artificial heart valves.** Design a one-way valve to be used as a replacement for a human heart valve. Outline a testing procedure for this valve. List all reference materials you use.

The heart is a four-chambered pump, and each chamber has an output valve. Disease sometimes affects the mitral valve of humans. When this occurs, it is sometimes advantageous to replace this valve with an artificial valve. Many types of valves, including pig heart valves, have been tried; none have been entirely satisfactory. The heart beats about 80 times per minute. The valve must withstand a back pressure of at least 200 millimeters of mercury. Artificial valves have been used on children as well as adults. With present valves, the survival rate after ten years is about 65%.

Problems with present valves are due to infection, rejection, wear, the slow rate of tissue in-growth, and thromboembolism (the obstruction of a blood vessel with a blood clot that has broken loose from its site of formation).

To market this new medical device, you will need approval from the Federal Food and Drug Administration (FDA). To gain this approval you will need, among other things, data collected from animal experiments and carefully controlled clinical trials on humans. As a part of the design of this valve, you must outline a testing procedure that will help you obtain FDA approval. Things you should consider include:

1. Biocompatibility: The surrounding tissues should not be adversely affected by implantation, nor should the degradation of the product be detrimental to the biological tissues.

2. Noncorrosiveness: The physiological fluids in the body correspond to a dilute saline solution, 0.9% by weight isotonic NaCl. For this reason, the implant materials must be able to resist corrosion that might be caused by this fluid. Also, if metals are used, caution must be taken to keep from setting up galvanic cells that will cause galvanic corrosion. Corrosion is not only harmful to the tissues by inflaming and irritating them, but it will ultimately lead to the failure of the implant.

3. Chemical stability and inertness are related to the corrosion-resistant properties of an implant. Even if corrosion is not present, the chemical inertness and stability come into play. For example, an implant may absorb body fluids that might change its properties.

4. Mechanical and physical compatibility require that the valve fit in its proper place in the heart. Consider the different sizes of valves necessary for children.

5. Thrombo-resistance is one of our major concerns. The implant material must permit the normal flow of blood without causing coagulation of the blood.

6. Long service implants must not degrade in the environment of the body. Some materials degrade, which leads to a reduced life of the product.

7. The valves must have the necessary strength to maintain their shape and serve their purpose.

8. Wear resistance must be considered. Low friction is desired. You do not want to produce potentially hazardous wear fragments.

9. Electrical characteristics, if any, should be inactive with respect to the surrounding tissues. The valve should not be adversely affected by microwave ovens.

10. If the valve will be exposed to magnetic fields, it may have to be built of nonferrous metals. Otherwise, the device might fail if the person were subjected to a high magnetic field.

11. The following numbers should be of interest to you: A heart valve costs $3000. The operation to install it costs $15,000 and maintenance

for the first year is about $100. The following numbers are for heart transplants, not heart valves, but they give insight to costs associated with open heart surgery. A heart transplant costs $85,000 plus $10,000 for drugs in the first year. Costs for the ensuing years will be $4500 for drugs and $10,000 for hospital and doctors' fees. About 2000 heart transplants are performed per year. While waiting for donors, some patients receive artificial hearts that can cost $300,000. [These numbers are courtesy of Jack Copeland (heart) and Sharron Schnider (valves).]

13. **Flywheel energy storage system.**   Power brownouts and blackouts have become common in many countries. Suppose a homeowner decides to remedy this problem by installing an electric motor/generator with a large flywheel. At night, when the electric load is low and power is cheaper, the unit will run as a motor, storing energy in the flywheel. During hot summer days, when the power from the utility might fail, the flywheel will run the generator and power the homeowner's electric load, which consists of air conditioners, electric ovens, clocks, etc.—about 10 kilowatts of load altogether. (Flywheels have also been suggested as a means for storing energy for buses and street cars.) You are asked to find out whether this idea has merit using space, weight, and cost as figures of merit.

The flywheel can have a maximum speed of 10,000 revolutions per minute. In delivering energy it may slow down to 50% of its maximum speed. It must supply 10 kilowatts for 10 hours. Consider only a simple flywheel having a rim one-tenth the radius of the wheel connected to the hub with light spokes. By keeping friction low, it is hoped that 85% of the total energy stored in the flywheel can be recovered as electric energy. One of your biggest design constraints is to make sure that the flywheel does not fly apart due to centrifugal forces. Assume that the material you are using has a tensile strength of 150,000 pounds per square inch.

In your failure analysis, discuss what happens if your flywheel breaks. Will it be a catastrophe?

14. **Postage scales.**   The cost of postage in the United States has been rising rapidly in the last few years. Unless some unforeseen technology or procedure is introduced, it appears that the cost of delivering a one-ounce first class letter may be two dollars by the year 2010. Some of the contributing reasons include:

1. a general inflationary trend in the economy,

2. a more rapid increase in labor costs in the service sector as compared with labor costs in other parts of the economy, and

3. the private sector paying the bills for all government and political mailings, which have increased markedly with the growth of government and increase in political activity in the U.S. in the past few years, particularly with the advent of computers that can produce address labels rapidly and at a low cost.

Attempts have been made to automate the post office because people do not want the boring jobs, but automation must be carefully applied. Otherwise, the result can be an increase in cost merely to appear modern.

Your task is to design, build, and test a postage scale that is inexpensive, convenient, and accurate. Dampening of the scale from the workers' hands will not be allowed. Letters must remain dry and unmutilated. No materials can be considered free of cost. Estimate the price you would have to pay for the materials to make 10,000 of these devices. The scale should be calibrated to give the postage required for any size first class letter up to 24 by 32 centimeters and any weight from zero to five ounces. For calibration purposes, it is convenient to know that 20 normal (3 by 5 inch) index cards weigh one ounce.

In solving this problem, you can use any principle or materials available, but the quality of your design will be measured by the final Trade-Off Figure of Merit (TFOP) defined as follows:

$$\text{TFOP} = \frac{100 - \text{Penalty}}{(\text{Cost})(\text{Time})} \ \sec^{-1}$$

or perhaps

$$\text{TFOP} = \frac{10^4 - 100\,\text{Penalty}}{(100 + \text{Cost})(\text{Time} + 10)} \ \sec^{-1}$$

where, Cost (in cents) is the cost of the materials used to build your scale; Time (in seconds) is the time needed to determine the postage required for five letters ranging in weight from 0 to 5 ounces; and Penalty (in cents) is the penalty for a lack of accuracy. This penalty is determined by comparing the postage your scale indicates with the correct value and summing the absolute differences for the five weighings. Which of the two above equations for the figures of merit do you think would be best? Why? Throughout most of this book our final Trade-Off Figures of Merit were linear combinations of the overall Performance Figure of Merit and the overall Utilization of Resources Figure of Merit. In this case, this combination is a ratio instead of a sum.

All systems will be verified publicly by the inventors, who will explain how they arrived at the Cost. Your Teaching Assistant may suggest a different value for Cost if some of the assumptions seem unreasonable. The values of Time and Penalty will be measured when the inventors weigh five "letters" provided by the instructor. All groups will have five minutes to set up their system and have it ready to begin the weighings. Any design that takes more than five minutes to set up will be disqualified.

15. **Electrical transmission line towers.** Electrical transmission line towers are often subjected to horizontal forces, such as transverse forces due to the wind blowing against ice covered conductors and longitudinal forces

due to broken conductors. Your task is to build a model of a tower designed to resist a horizontal force of 89 newtons applied 50 centimeters above the ground. The following technology limitations must be observed:

1. The tower must have a base that will fit into the clamp and support shown in Figure 8.3.

2. A horizontal force of 89 newtons will be applied to the structure by means of a string attached to a one inch diameter washer. Your system must accommodate this string and washer. You may not tie knots in the string.

3. You are limited to the use of balsa wood and Elmer's® white glue. No pins, wires, string, or other materials may be used.

The designs will be judged not only on strength-to-weight ratio, but also on the aesthetic appearance of the structure and the apparent ease in constructing a full-scale counterpart. In this respect, it should be remembered that we are building models of full-size transmission line towers that could visually dominate the landscape.

To test your system, a horizontal force of 89 newtons will be slowly applied with a string and pulley system. Once the force reaches 89 newtons, it will be maintained for ten seconds. Then the force will be slowly increased until the tower breaks. The breaking force will be recorded; it will be used to help calculate the overall Performance Figure of Merit.

The overall Performance Figure of Merit (IFOP) will be based on the following figures of merit:

1. Weight Supported Before Failure,
2. Apparent Ease of Fabrication of Full-Scale Counterparts, and
3. Appearance.

The overall Utilization of Resources Figure of Merit (UFOP) will be based on the following figures of merit:

1. Amount of Glue and
2. Total Weight of the Tower.

The final Trade-Off Figure of Merit will be defined as

$$TFOP = \frac{IFOP}{UFOP}$$

An important part of this project is deciding the appropriate weights for the figures of merit. Before construction begins, each class should decide on these weights. You might even decide to omit some of our figures of merit and add some of your own.

Some instruction should be given in class regarding basic structural analysis. You should remember that when John Roebling designed the

Brooklyn Bridge, his knowledge of basic engineering principles (statics, dynamics, hydraulics, and structural analysis) was not much more than yours is now.

Your tower should be designed to withstand unexpected events. For example, if the Teaching Assistant applying the load to your tower sneezes, or if an earthquake shakes her, she might apply transverse forces. It would be to your advantage if your tower survived such an event and could be tested

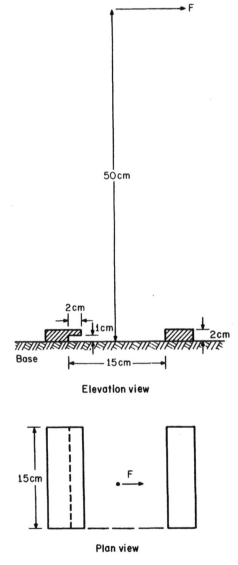

Figure 8.3 The clamp that will hold your tower.

later. Your documentation should explicitly describe how you have taken into account such examples of Murphy's law.

16. **The Lego® bridge.** Build the Popsicle® stick bridge or the electrical transmission line tower out of Lego bricks instead of wood. Or, use an advanced Lego kit, a Fishertechnik® kit, or some other kit provided by your instructor to make a machine that can move three meters (or ten feet) in three minutes. Develop a test requirement that can be used to judge the system.

17. **Crane speed control.** A construction elevator is used to haul material up and down a high-rise building under construction. The problem with the system is that the hoist (with a maximum load of 100 kilograms) elevates slower than is desired. Conversely, when the hoist carries debris back down, it goes too fast. If a 100-kilogram load is brought down from the top of the building, which is 50 meters above ground, the hoist can attain a speed of 25 meters per second. The motor must be controlled in such a way as to provide a braking torque.

We want to provide an automatic control system for the motor, as shown in Figure 8.4. A tachometer is attached to the motor shaft to measure the speed. The tachometer produces a voltage proportional to the shaft rpm:

$$v = \frac{N}{75} \frac{\text{volts}}{\text{rpm}}$$

A reference voltage, V_{ref}, is selected by the operator in the range −24 to +24 volts. The error between the tachometer voltage and V_{ref} is amplified by a power amplifier with an output of A times the error. We would like to select the amplifier gain A so that the speed error does not exceed 5%. The motor nameplate reads:

Maximum speed	1800 rpm
Rated voltage	240 volts DC
Rated current	16 amperes
Field current	1.5 amperes
Moment of inertia	0.01 kg·m²
Armature resistance	0.5 Ω

The steady-state motor characteristics are given by:

T = motor torque
    = $20I$ newton-meters, where
$I \doteq$ current in amps,
E = terminal voltage
    = $0.133N + RI$ volts, where
N = speed in rpm and
R = armature resistance.

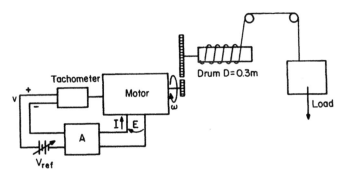

Figure 8.4 A crane motor system.

There is a 10:1 gear reduction between the motor and the drum. The load can vary from 0 to 100 kilograms.

18. **Oil-filled cylinder.** This is a "black box" problem, in which you are to determine a physical property of a complex object by using simple laboratory devices and basic principles. You will be given a metal cylinder closed at both ends. All you know about the cylinder is its length, its outside diameter, and its contents—it is filled with oil.

Your task is to experimentally determine the polar moment of inertia of the steel container only, excluding the oil. You should discuss the probable accuracy of your results and perform an independent approximate check on your answer.

19. **Scheduling hospital appointments.** If you are able to gain the cooperation of some hospital administrators, you might be able to design a scheduling system for the appointments of hospital outpatients. The following description is specific to an oncology center. Typically, scheduling is done at five separate locations:

1. Receptionist's desk: The receptionist schedules new patients for initial and follow-up examinations by a given doctor.

2. Operating room: An operating room is available to the center one day each week. The operating room is used for patients receiving radiation implants as a means of cancer therapy. Doctors requiring the use of the operating room must request scheduled time.

3. Linear accelerators: There are four linear accelerators that are used for patient treatment and for research by the laboratories within the center. Each machine has a schedule book. Scheduling is done by the technician who operates the machine. The patient's doctor must be present for the initial setup of the patient on the machine.

4. Simulator: Before treatment on a linear accelerator, a patient is viewed by a simulator that uses X-rays to determine the exact location

and extent of disease. The patient's appointment time is recorded in a separate schedule book by the simulator technician. The patient's doctor must be present for simulation.

5. Secretary: Each of the doctor's secretaries keeps separate schedules. Each schedule contains the doctor's appointments, meetings, days off, travel plans, etc.

A lot of time and money is lost if these schedules are not coordinated. Design a system to coordinate these schedules.

20. **Powering a light bulb.** Design and build a device to illuminate a light bulb. Your device may not consist of commercially available batteries, nor may it use 115 volt AC power. Figures of merit that can be used to evaluate alternatives include: Historical Age of the Device, Cost of Your Components, Length of Time It Powers the Light Bulb, and Brightness of the Bulb. Brightness can be measured with a photometer, or it can be assessed by class vote. Assigning weights and combining these figures of merit into one overall figure of merit is an important part of this project. In your preliminary designs you might consult books such as R. S. Kirby et al., *Engineering in History*, McGraw-Hill, New York, 1956 or Dover, New York, 1990.

21. **Combinational logic.** Assume that you are working on a text processor for the Hawaiian language. Of course, you will eventually have a spelling checker and other extras. But, just as a simple debugger, design and build a combinational logic circuit that will detect the letters of the Hawaiian alphabet. The letters will be stored as ASCII characters, as shown in the following table, where A through G are designations for binary bits, with A being the most significant bit and G being the least significant bit. For example, the letter "Z" is represented by $1011010 = A\bar{B}CD\bar{E}F\bar{G}$. In the Control columns, the abbreviation CR stands for carriage return and LF stands for line feed. The other abbreviations are discussed in many digital systems design books, such as J. P. Hayes, *Digital System Design and Microprocessors*, McGraw-Hill, New York, 1984, p. 389.

Your circuit will have seven inputs (use DIP switches) and one output (use an LED). When the settings of the input switches correspond to ASCII characters of the Hawaiian alphabet the LED should turn on. The best circuits will require the fewest integrated circuits and the fewest interconnecting wires. Discuss the alternative designs that you considered in your report. Also, be sure to include your circuit diagram and your pin-to-pin wire list. You may buy any integrated circuits that you wish to use. This project requires a few weeks of lectures on combinational logic.

How do you plan to test your circuit so as to convince the Teaching Assistant that it works? A primitive testing technique would be to input two correct combinations and two incorrect combinations and see if the outputs are correct in all cases. A more exhaustive testing plan would be to apply all 128 possible input combinations and record the outputs. However, we doubt

The ASCII code							
The least significant bits	Control		Numbers and symbols		Upper case		Lower case	
	The most significant bits, ABC							
DEFG	000	001	010	011	100	101	110	111
0000	NUL	DLE	space	0	@	P	'	p
0001	SOH	DC1	!	1	A	Q	a	q
0010	STX	DC2	"	2	B	R	b	r
0011	ETX	DC3	#	3	C	S	c	s
0100	EOT	DC4	$	4	D	T	d	t
0101	ENQ	NAK	%	5	E	U	e	u
0110	ACK	SYN	&	6	F	V	f	v
0111	BEL	ETB	'	7	G	W	g	w
1000	BS	CAN	(	8	H	X	h	x
1001	HT	EM	)	9	I	Y	i	y
1010	LF	SUB	*	:	J	Z	j	z
1011	VT	ESC	+	;	K	[	k	{
1100	FF	FS	,	<	L	\	l	|
1101	CR	GS	-	=	M	]	m	}
1110	SO	RS	.	>	N	^	n	~
1111	SI	US	/	?	O	_	o	delete

that many Teaching Assistants would have the patience to sit through such a test. Suggest the design of a system that could test your system for malfunctions, but do not build it. That is, any time your system is not performing its task properly, it could be programmed to do self testing.

22. **A digital clock.** Write an assembly language (or some other language) program that will simulate a digital clock. The clock should display minutes and seconds in decimal numbers. The format of the output should be $m_1m_0 : s_1s_0$, where m_1m_0 is a two-digit decimal number between 0 and 59 representing minutes and s_1s_0 represents seconds. Your program should generate the time for seconds using a loop that counts down from some large number to zero or by performing some large number of other operations that will delay time. (We do not expect it to be very accurate.) After the clock is incremented, the program should replace the current display with the new time (just as a digital watch does). At first, your waste-time loop should be long enough to clearly show the lab instructor the first transition from seconds to m_0. After this, the waste-time loop should be shortened—to make the clock run faster—so that the transition from 59:59 to 00:00 can be seen by the lab instructor. Design, but do not build, a system to verify that your clock is performing correctly.

Who's got the digital watch? Ask two people what time it is. The person with the analog watch will say, "It's a quarter to five," and the person with the digital watch will say, "It's four forty-four." The digital watch owner's response fallaciously implies greater accuracy. He probably set his watch last month using the clock on his stove, whereas the guy with the analog watch may have set his watch that morning using radio station WWV. On a flight across the international date line, a pilot was reported to have said, "Here in Samoa it is 4:30 on Tuesday, but at our destination in the Fiji Islands, it is 3:30 on Wednesday. So when we land, those of you with analog watches should turn your watches back one hour. For those of you with digital watches, God help you. As for me, I prefer a sundial."

23. **The SIE Railroad.** Design and build the three controllers mentioned in Chapter 6 for the Systems and Industrial Engineering Railroad. In your first design, use integrated circuits and wires to form a hard-wired controller. Build this on the protoboard and control the trains with it. In your second design, write a high-level (i.e., Pascal) program to do the same job. In your third design, write an assembly language program to do the same job. In your final report, you must compare and contrast these disparate technologies. Include a discussion of cost per unit, capital costs, number of units to be built, time of construction, ease of construction, ease of maintenance, ease of modification, ease of modification by someone other than the designer, reliability, debugging, and testing methods.

For ease of construction you will probably assume that no two sensors can be activated simultaneously. However, you should analyze your systems and explain what each of them would do if there were simultaneous inputs.

How will you test your system? One obvious way is to use four switches to represent the input sensors and two LEDs to represent the output power controls. Apply the input trajectories, as shown in Chapter 6, and record the output trajectories. Design, but do not build, an appendage to your system that could test the whole system for proper operation any time a Built-In Test Signal (BITS) is presented.

24. **Don't flatten your soda pop.** Do you get annoyed when you pour your soda pop over ice and all the bubbles come out? Let me suggest a simple alternative: Wash your ice cubes first. First, fill your glass with ice. Then, cover the ice with water and quickly put your fingers over the top of the glass and pour out the water. Now when you pour in the soda pop, you will not lose the bubbles. Why does this work? Do you have dirty ice? Describe this method as a series of functions with inputs and outputs.

25. **An electric car.** Due to government regulations, automakers are now designing and building electric cars. Some are battery powered, and some are hybrids that use both electric and petroleum-based motors. Write the Input/Output Performance Figures of Merit and the Utilization of Resources Figures of Merit for a battery-powered car. Discuss which figures of

merit must be traded-off against others. For an excellent article about the systems engineering of GM's electric car, see David H. Freedman, "Batteries Included," *Discover*, March 1992, pp. 90–98.

26. **Index Card Towers.** Other popular projects involve building structures with cardboard index cards. The only materials allowed are 3 by 5 inch index cards and paper clips. Two suggested projects are:

1. design and build the tallest tower that will support one red brick and

2. design and build the structure that will support the most red bricks at least three inches off the ground.

Of course, the weight of the red bricks must be specified, and specifying a maximum number of index cards is advisable. Figures of merit must also be designed. Some reasonable ones include:

$$FOM_1 = \frac{\text{Height of the tower in centimeters}}{w_1 \times \text{Number of index cards} + w_2 \times \text{Number of paper clips}}$$

$$FOM_2 = w_1 \times \text{Number of bricks supported} - w_2 \times \text{Number of index cards} - w_3 \times \text{Number of paper clips}$$

In addition to charging for materials (cards and paper clips), alternative figures of merit charge for manufacturing processes (folds in the cards and bends in the paper clips).

Tools, such as rubber bands, may be used in the construction if they are removed at the end of construction. Alignment is critical. A flashlight is helpful in establishing absolute verticality.

8.2 Other projects

1. Design a system to automatically connect the battery of an electric car to a power source when it is parked in its garage. At a minimum, your system should have contacts, an on/off switch, a fuse or circuit breaker, and a good-contact indicator.

2. Design a device to prevent the theft of helmets left on motorcycles.

3. Design a system to keep the engine of an automobile warm overnight. What failures could occur in your device? What are the consequences of these failures?

4. Discuss the differences in two law enforcement systems. One of the systems specifically excludes firearms, wheras the other allows them.

5. Design a device to alert drivers who might be falling asleep. Be sure that a failure in your system would not go undetected. How would you test your system?

6. Design a device that will allow my dog, but no other animal or human, to enter and leave my house. She is a big dog.

7. Design a household water conservation system.

8. Design a system to provide food and water for a dog while its owners are away for the weekend. If your device were to fail (due to a lightning strike, power failure, etc.), would it be better to overfeed or underfeed the dog?

9. Discuss the role of alchemists in defining a technology for a project that has the input/output specification of turning lead into gold.

10. Design a device to be installed in a car or a truck that will show a traffic inspector that the speed limit has been exceeded. The design of the device must take into account the fact that honesty is not an inherent property of human beings. How will you test your system?

11. Design a device that will inform the handlers of a fragile package that it is suffering undesirable shocks during transportation. Will your system be one-shot (like a flash bulb) or reusable (like a strobe light)? Design a test program for both types.

13. Design an automobile instrument panel. Include gauges for water temperature and level, transmission fluid level, the need for an oil change, etc. Show power and sensor connections. Is your system fail-safe if an indicator light burns out?

14. Design a system for improving an automobile's traction on ice. Default systems include snow tires (with or without studs), tire chains, cable chains, and four-wheel drive. How will you test your system?

8.3 *Large Projects*

Real projects must have a real customer. For one thing, this allows students to discover for themselves that customers seldom understand their own needs. However, it is hard to get experts in the manufacturing industry to donate their time so that students can learn to design systems. We have found that agencies sponsored by the United Way are often willing to participate. They usually have low budgets and cannot afford to hire engineers, but they appreciate help from student engineers. Often students design systems that are quite helpful to them. We have had students design systems for managing juvenile delinquents, managing laundry at a retirement home, scheduling nurses, controlling a telephone information hotline, controlling books in a library, and allocating hiking permits in national parks.

We have also found that students with part-time employment can find valuable projects in their work environment. Parents and siblings also make ideal customers. Using such sources, our students have designed systems to:

- select cookware in a department store,

- teach and help medical professionals identify chromosomal abnormalities in infants,
- help with the diagnosis and prognosis of children who begin to stutter,
- help an emergency room physician diagnose ophthalmological problems,
- help an ophthalmologist diagnose disease using fundus photos,
- help a pediatrician identify the cause of a rash on a child,
- help evaluate student study plans,
- help students use a personal computer laboratory,
- help students use DOS- and UNIX-based personal computer laboratories,
- help install 4.2BSD Unix on a VAX computer,
- give homeowners advice to help them save money on their electric bills,
- regulate human use of the Grand Canyon National Park,
- find the best scheduling rule for a job shop,
- provide advice on modifying a Volkswagen 1600cc engine to optimize its performance,
- help design solar energy home remodeling plans,
- aid a biomedical engineer in selecting BMDP statistical analysis programs,
- help design rockbolt support systems for coal mines,
- assist librarians in assigning subject call numbers to books using the Library of Congress system of cataloging,
- help construction engineers repair roads,
- help diagnose automobile failures,
- help a person run a four-color printing press,
- help lawyers plea bargain narcotics cases,
- help relocate a University Physical Resources Center,
- help the Indian Health Service reduce the incidence and severity of alcoholism among Native Americans,
- schedule bartenders, and
- help school administrators mainstream handicapped students.

Our students have also designed

- a personal computer laboratory,
- personal computer local area networks,
- an automated mail delivery system,
- a scheduler for operations in a machine shop,
- a vadose zone monitor to alert for possible groundwater pollution,
- a troubleshooter for a wire-bond system,
- a scheduler for hospital outpatient appointments,
- a system to put sound on film,
- a system to reduce obstetrical risk,
- a system to discover the causes of failure in RS-232 connections between a terminal and a computer,
- a digital image management system for a radiology department,
- a rural transportation system for Pinal County, and
- an agricultural information system for Ecuador.

Designing tests to prove that a system does what it was supposed to do is an important aspect of system design. We have shown only a few techniques for designing tests, since further discussion is not within the scope of this textbook. Bahill (1991) presents several more techniques for testing systems; these tests were applied to many of the systems mentioned in the preceding paragraphs.

appendix

Computer programs for scoring functions

A floppy disk containing an interactive program that plots the scoring functions of Figure 4.2 is available from the authors. It was written by Dave Voorhees using Lotus 1-2-3® and runs on IBM-compatible computers. The values for scoring functions may also be computed using the C language program listed below.

```
#include <stdio.h>
main()
    {
    double ssf(), score, a;

    /* example of a one-sided scoring function */
    printf ("One sided scoring function\n");
    for (a=0.0; a<=5.0; a+= 0.25)
        {
        /* lower = 0.0, baseline = 1.0, upper = 5.0,
        slope = 0.4, fom = a */
        score = ssf( 0.0, 1.0, 5.0, 0.4, a);
        printf("a=%f score=%f\n",a,score);
        }

    /* example of a two-sided scoring function */
    printf("\nTwo sided scoring function\n");
    for (a=0.0; a<=5.0; a+= 0.25)
        {
        /* lower = 0.0, lbaseline = 1.0, opt = 5.0,
            slope = 0.5, fom = a */
        score = ssf( 0.0, 1.0, 5.0, 0.5, a);
        printf("a=%f score=%f\n",a,score);
        }
    for (a=5.0; a<=10.0; a+= 0.25)
        {
        /* opt = 5.0, ubaseline = 7.0, upper = 10.0,
```

```
        slope = -0.7, fom = a */
    score = ssf( 5.0, 7.0, 10.0, -0.7, a);
    printf("a=%f score=%f\n",a,score);
    }
 }

/* Standard Scoring Function */
/* l => lower threshold      */
/* b => baseline             */
/* u => upper threshold      */
/* s => slope at baseline    */
/* v => figure of merit      */
double ssf(l,b,u,s,v)
double l,b,u,s,v;
  {
  double score, ssf1();
  int negslope;

  /* score => calculated score based on input */
  /* negslope => flag indicating negative
                   slope */
  if (s < 0.0) {
    s = -s;
    negslope = 1;
    }
  else
    negslope = 0;

  /* figure of merit is above the baseline */
  if (v > b) {
    score =  1.0 - (ssf1(2.0*b-u,b,s,2.0*b-v));
    }

  /* figure of merit is below the baseline */
  else
    score = ssf1(l,b,s,v)
  if (negslope == 1)
    score = 1.0 - score;
  return (score);
  }

/* this routine does the math */
double ssf1(l,b,s,v)
double l,b,s,v;
  {
/* l => lower threshold        */
```

```
/* b => baseline                 */
/* s => slope at baseline        */
/* v => figure of merit          */
/* x => power of function        */
/* BIGNUM => parameter based on computer limits
              (999.0 is okay) */
  static float BIGNUM = 999.0;
  double x, pow();
  if (v <= l) {
    return (0.0);
    }
  else
    {
    x = 2.0*s*(b+v-2.0*l);
    if (x > BIGNUM)
        x = BIGNUM;
    return ( 1.0/ (1.0 + pow(((b-l)/(v-l)),x)) );
    }
  }
```

Results of running the above program:

One-sided scoring function

```
a=0.000000 score=0.000000
a=0.250000 score=0.200000
a=0.500000 score=0.303270
a=0.750000 score=0.400651
a=1.000000 score=0.500000
a=1.250000 score=0.598721
a=1.500000 score=0.690229
a=1.750000 score=0.769290
a=2.000000 score=0.833553
a=2.250000 score=0.883226
a=2.500000 score=0.920123
a=2.750000 score=0.946689
a=3.000000 score=0.965347
a=3.250000 score=0.978177
a=3.500000 score=0.986818
a=3.750000 score=0.992499
a=4.000000 score=0.996109
a=4.250000 score=0.998276
a=4.500000 score=0.999439
a=4.750000 score=0.999919
a=5.000000 score=1.000000
```

Two-sided scoring function

```
a=0.000000  score=0.000000
a=0.250000  score=0.150221
a=0.500000  score=0.261204
a=0.750000  score=0.376732
a=1.000000  score=0.500000
a=1.250000  score=0.622500
a=1.500000  score=0.731351
a=1.750000  score=0.818376
a=2.000000  score=0.882236
a=2.250000  score=0.926162
a=2.500000  score=0.954999
a=2.750000  score=0.973300
a=3.000000  score=0.984615
a=3.250000  score=0.991451
a=3.500000  score=0.995479
a=3.750000  score=0.997777
a=4.000000  score=0.999024
a=4.250000  score=0.999648
a=4.500000  score=0.999914
a=4.750000  score=0.999992
a=5.000000  score=1.000000
a=5.000000  score=1.000000
a=5.250000  score=0.998572
a=5.500000  score=0.992248
a=5.750000  score=0.977603
a=6.000000  score=0.948398
a=6.250000  score=0.894591
a=6.500000  score=0.803709
a=6.750000  score=0.668418
a=7.000000  score=0.500000
a=7.250000  score=0.331714
a=7.500000  score=0.197202
a=7.750000  score=0.107699
a=8.000000  score=0.055292
a=8.250000  score=0.027006
a=8.500000  score=0.012532
a=8.750000  score=0.005437
a=9.000000  score=0.002124
a=9.250000  score=0.000690
a=9.500000  score=0.000154
a=9.750000  score=0.000012
a=10.000000 score=0.000000
```

Bibliography

Akao, Y., ed. (1990) *Quality Function Deployment: Integrating Customer Requirements into Product Design*. Productivity Press, Cambridge, MA.

American Supplier Institute (1988) *R&M 2000 VRP Conference Notes*. American Supplier Institute, Dearborn, MI.

Aslaksen, E. and R. Belcher (1992) *Systems Engineering*. Prentice-Hall, Englewood Cliffs, NJ.

Bahill, A. T. (1981) *Bioengineering: Biomedical, Medical and Clinical Engineering*. Prentice-Hall, Englewood Cliffs, NJ.

Bahill, A. T. (1991) *Verifying and Validating Personal Computer-Based Expert Systems*. Prentice-Hall, Englewood Cliffs, NJ.

Bahill, A. T. and W. J. Karnavas (1991) The Ideal Baseball Bat. *New Scientist*. Vol. 130, No. 1763, pp. 26–31.

Blanchard, B. S. and W. J. Fabrycky (1990) *Systems Engineering and Analysis*. Prentice-Hall, Englewood Cliffs, NJ.

Bossert, J. L. (1991) *Quality Function Deployment: A Practioner's Approach*. ASQC Quality Press, Milwaukee, WI.

Clausing, D. (1990) Concurrent Engineering. Paper presented at the Concurrent Engineering Design Clinic, June 7, Ypsilanti, MI.

Clausing, D. and E. Pugh (1991) In *Proceedings of the Design Productivity International Conference*, Feb. 3–9, Hawaii. Design Productivity Center, University of Missouri, Rolla, MO.

Clements, R. R. (1989) *Mathematical Modeling: A Case Study Approach*. Cambridge University Press, New York.

Concurrent Engineering (1991) *IEEE Spectrum*. Vol. 28, No. 7, pp. 22–37.

Conklin, E. J. (1987) Hypertext: An Introduction and Survey. *IEEE Computer*. Vol. 20, No. 9, pp. 17–41.

Department of Defense (1985) Critical Path Templates for Transition from Development to Production, DOD 4245.7-M. Washington, D.C.

Harrington, H. J. (1987) *The Improvement Process*. Quality Press, Div. of McGraw-Hill, New York.

Jeffreys, W. H. and J. O. Berger (1992) Ockham's Razor and Bayesian Analysis. *American Scientist*, Vol. 80, pp. 64–72.

Karnavas, W. J., P. Sanchez, and A. T. Bahill (1993) Sensitivity Analyses of Continuous and Discrete Systems in the Time and Frequency Domains. *IEEE Transactions on Systems, Man and Cybernetics*, SMC-23.

King, B. (1989) *Better Designs in Half the Time, Implementing QFD Quality Function Deployment in America.* Goal/QPC, Methuen, MA.

Kerzner, H. (1989) *Project Management: A Systems Approach to Planning, Scheduling, and Controlling.* Van Nostrand Reinhold, New York.

Lake, J. G. (1991) Concurrent Engineering: A New Initiative. *Program Manager.* Sept.–Oct., pp. 18–25.

MacDonald, C. D. R. (1987) *Intuition to Implementation.* Prentice-Hall, Englewood Cliffs, NJ.

QFD/Capture User's Manual (1990) International TechneGroup, Milford, OH.

Re Velle, J. B. (1988) *The New Quality Technology: An Introduction to Quality Function Deployment (QFD) and the Taguchi Methods.* Hughes Aircraft Co., Los Angeles.

Simon, H. A. (1962) The Architecture of Complexity. *Proceedings of the American Philosophical Society,* Vol. 106, pp. 467–482.

Szidarovszky, F. and A. T. Bahill (1992) *Linear Systems Theory.* CRC Press, Boca Raton, FL.

Szidarovszky, F., M. E. Gershon, and L. Duckstein (1986) *Techniques for Multiobjective Decision Making in Systems Management.* Elsevier, Amsterdam.

Transactions of the Symposium on Quality Function Deployment (1989, 1990, 1991) Novi, MI, Goal/QPC, Methuen, MA, and the American Supplier Institute, Dearborn, MI.

U.S. Navy (1986) *Best Practices.* NAVSO P-6071.

Ver Planck, D. W. and B. R. Teare (1954) *Engineering Analysis: An Introduction to the Professional Method.* Wiley, New York.

Watts, R. G. and A. T. Bahill (1990) *Keep Your Eye on the Ball: The Science and Folklore of Baseball.* W. H. Freeman, New York.

Wymore, A. W. (1976) *Systems Engineering Methodology for Interdisciplinary Teams.* John Wiley & Sons, New York.

Wymore, A. W. (in press) *Model-Based Systems Engineering.* CRC Press, Boca Raton, FL.

Yakowitz, S. J. and F. Szidarovszky (1989) *An Introduction to Numerical Computations.* Macmillan, New York.

Index